PROMETHEUS'S BROTHER

Thomas Pornin

Prometheus's Brother

Gods, Science, and Camels

Cover illustration: terracotta figurine of a camel carrying transport amphorae, Egypt, late 2nd or early 3rd century AD; currently on display in the Metropolitan Museum of Art (New York), gift of Mrs. Lucy W. Drexel, 1889. The photograph was made available under the MET's OASC initiative (Open Access for Scholarly Content); see: http://www.metmuseum.org/

Title page illustration: *Chimera*, a mythological beast assembled from parts of several different animals, and breathing fire. This drawing was originally made by Pearson Scott Foresman, who released it into the Public Domain.

Additional copyright notices for the source material for the illustrations are given in the "Acknowledgements" section.

Printed by CreateSpace.

ISBN-13: 978-1533126177

ISBN-10: 1533126178

Contents

Foreword

Most people don't give more than a cursory glance to the world we live in today. They understand that history happened giving us all manner of interesting fossils and Discovery Channel programmes, but how it happened, and what that means to modern culture, are not forefront in the mind of the average man or woman on the street. In the past, the church told the majority what to think. Nowadays television has taken over that role. I'd like to think that the majority of the Millenial generation will know of Charles Darwin, but am I confident that they will get past the common misconceptions of natural selection and the YouTube-worthy contenders for the Darwin Awards?

Not really.

For those of us with a curious mind, however, the how it happened, and the why, and all the other questions, are almost more important than the end result. As a child I was provided with the Encyclopaedia Britannica, National Geographic, New Scientist and so on, and was fascinated by every discovery and every new understanding. This founded a lifelong love of the workings of the Earth and the universe, and a desire to find out more.

I have the pleasure of knowing a large number of inquisitive individuals with a similar curiosity. It's inevitable in my line of work, which attracts a high percentage of people who wish to dismantle what they see, the better to understand it, fix it and improve it. This is the mindset of the Engineer, the Scientist, the Hacker, and Thomas Pornin is almost the epitome of these, combining diligent research with appropriate levels of irreverent speculation.

In the few years I have known Thomas I have been continually impressed and entertained by his approach to almost any topic you can think of. In sharing his experience he tackles surface questions deftly, and with a certain sarcastic humour then looks at the implied framework underpinning the question in order to improve on the question itself, provide alternate answers, and to discuss why the question may not be the important thing to look at after all.

In Prometheus's Brother, Thomas looks even deeper, taking you on a journey through time from creation to the present day, and along the way looks at philosophy, cultural and legal implications of the human approaches to identifying and understanding animals, and the ramifications of humans playing god in the sandbox that is Earth.

Along the way, you'll learn a lot about camels, clay, and catastrophe, and every chapter will leave you educated in areas you may not have expected. And I guarantee you'll be pondering the likely winners and losers of the 6th Extinction long after you close this book.

Rory Alsop, June 2016, Scotland

Introduction

This book is about the history of an idea. That idea is, roughly speaking, the way we, humans, look at the diversity of lifeforms in Nature: we insist, sometimes against all evidence to the contrary, on classifying them into separate, neatly-defined categories, that we call species; we then make moral choices based on this classification. This leads to remarkable phenomena, such as huge efforts at conserving some species that are on the brink of extinction, while, at the same time, the very notion of species is increasingly harder to pinpoint in a scientific way.

My purpose is to explore that old idea, which survived more than two millennia of religion, philosophy, science and politics, and still drives our collective decisions and actions with regards to the ecosystems we live in. As such, my prose constantly jumps between these themes in a breathless attempt at describing enough context for the narrative to make sense. This is actually the way I think, and I am quite sure there is a psychiatrically adequate term to describe such mental vagrancy. I hope that it will not make your reading too uncomfortable.

In all this text, I never endeavour to convince you of anything; my goal is to present a collection of facts and possible links between them, so that you may do your own thinking. On any subject that triggers your interest, you should read several, many books, and compare the different points of view. At the end of this work, you will find a short bibliography that presents a few extra reading suggestions.

The story I relate is intimately linked with the rise of modern science, and in particular evolutionary biology, which happened mostly in Western Europe. This leads to a seemingly eurocentric point of view. I do not want to imply that other areas and peoples on this planet have been an intellectual desert; only that I did not take the time to learn, understand, and describe how natural history has been and is perceived in other cultures. I suspect it would require many more pages, probably thousands. I tried, with questionable success, to keep the scope of my discourse within reasonable limits.

If you reach the end of this book and then find that you either learned something, acquired some new "food for thought", or were simply satisfactorily entertained, then my efforts will not have gone in vain.

❧ 1 ❧

The Gods of Blunder

Prometheus is a mythological superstar. Though he appears in only a handful of myths and stories, his actions are decisive and a lot more meaningful and rife with symbols than the endless stream of romantic encounters of most Olympian gods. While Zeus, Aphrodite, Ares, Athena and their pals seem intent on begetting children to each other (and to mortal women), Prometheus has a much more serious curriculum vitae. Through the works of ancient poets, we do not see him aimlessly frolicking; instead, Prometheus is described as both creator and champion of Mankind. His opposition to Zeus, his tricks, his theft of the Divine Fire, and his harsh punishment make him a powerful and tragic figure, the perfect embodiment of Men's efforts at refusing their subservient status and at rivalling gods. Correspondingly, Prometheus has been used and reused as inspiration in many works of art and philosophy, especially as a metaphor for scientific endeavours and the innate tendency of Men to try to understand their world, crack open its secrets, reveal its mysteries, and more generally poke at things and fiddle with them.

This book is not about Prometheus. Instead, it concentrates more on an idea, a concept which is most concisely exposed in a story that involves Prometheus's brother, the much less famous Epimetheus. However, in order to recall that story, we must get some context about Prometheus's family.

It is important to make a few points clear about myths. In our modern world, a myth is at best a childish misconception, a lie, a falsehood that must be debunked. Calling a story a myth is strongly asserting that it contains nothing true. This meaning of "myth" is recent; in the Mediterranean and European area, it basically comes from the early christian efforts at converting pagans. Before the onset of Christianity, in particular in pre-Roman ancient Greece, myths had a much better reputation and were deemed to be all "true". It must be noted that concepts and ways of thinking evolve over time. In the days of Hesiod and Plato, our two main sources for Prometheus myths, monotheism had not been invented yet: there was no notion of a "fake god". For people at that time, when thinking about the divine world (as opposed to the world of Men, where objects can be touched and events directly witnessed), any story was true as long as it had been rhymed about by some poet.

Strange as it may be to our modern minds, in 500 BC, nobody would have understood the difference between "my god is stronger than yours" and "your god does not exist". The spiritual world, where deities reside, simply operated under a distinct modality from the mundane, tangible world. People found it only natural to use myths as the foundation for elaborate thinking. We will see that Plato, one of the finest philosophers of his era, does exactly that.

The consequence is that Greek myths have been influential themes for the development of philosophy and science throughout Middle Ages and later eras. This holds even though the wide corpus of Greek myths has never been unified, and stories are inconsistent with each other. Ancient Greece had no pope; there was no ultimate religious hierarchy that would ponder the veracity of any myth and establish a definitive canon[1]. Contradictions between texts do not diminish their importance in that respect. This means that when we study myths, we are really looking at the ancestors of our modern *ideas*.

That being said, let's see what stories say about Prometheus and his siblings. Surviving sources include mostly Hesiod's *Theogony* and *Works And Days*, and Plato's *Protagoras*. Hesiod may have lived at some time in the late 8th century or early 7th century BC; we don't really know for sure. The *Theogony* is an epic poem that aims at being a comprehensive genealogy of all gods; the *Works And Days* is a didactic almanac in which myths, in particular the one about Prometheus and Pandora, are recalled for their pedagogical value as illustration of Hesiod's notions of justice. Both texts were first composed and transmitted orally; indeed, literacy rates were abysmal in ancient Greece at that time.

This is, in itself, a very interesting and highly debated piece of history: at the end of the period traditionally called the "Bronze Age", around 1100 BC, almost all brilliant civilisations in the Aegean and Eastern Mediterranean collapsed. The exact causes are disputed, and appear to be complex, but the phenomenon was extensive. In Greece, in particular, urban sites were abandoned and populations reverted to nomadic pastoralism for

---

[1]A few centuries later, the Roman Republic would annex Greece; and the Republic had a *Pontifex Maximus*, by definition highest priest of all cults – since Augustus, this office would be held by the Emperor. However, this charge mostly entailed accountancy and legal management of marriages and successions, rather than maintaining storyline consistency.

about three centuries; that period is called the "Dark Ages". Since no written artefacts corresponding to the Dark Ages were ever found in Greece, archaeologists tend to conclude that these nomadic people were all illiterate. At the end of the Dark Ages, writing appears again, but with a different alphabet. At the end of the Bronze Age, Greek centres of culture, in particular Mycenae (the city ruled by Agamemnon in the Iliad), would use Minoan "linear B" syllabic glyphs, imported from Crete, to write down their language, which was an ancient form of Greek. Three hundred years later, when literacy was again on the rise, Greeks would use a completely distinct set of signs, alphabetic instead of syllabic, borrowed from Phoenicia. The compelling conclusion is that at the time of Hesiod, almost nobody would have been able to read any written poem, so it would have made little business sense for Hesiod to actually write, if he even knew how to do that. A poet like Hesiod would earn his living by composing poems in his head, then roaming from village to village to declaim his creations. Texts would survive by being learned and repeated by apprentices.

Plato lived about three centuries later, in the city of Athens, that was politically, militarily and culturally dominant over most of the Aegean area at that time. This is also the epoch when it became fashionable again to write things down, and Plato did just that. Most of Plato's production is about his former master Socrates and his teachings, which were strictly oral. Socrates left no text at all, and (according to Plato) he was in fact quite hostile to the whole notion of writing, because he believed it to be detrimental to the memory, and might induce the reader to acquire the appearance of knowledge without bothering to think. To our collective philosophical delight, Plato elected not to follow his master's steps in that respect, and he wrote a lot of things. Plato's *Protagoras* is a philosophical dialogue between Socrates and an unnamed companion, in which Socrates relates an earlier conversation he had with Protagoras, a major philosopher of the school known as Sophists[2]. In that nested dialogue, Protagoras endeavours to demonstrate his point to an unconvinced Socrates by starting with the recalling of a myth about Prometheus and Epimetheus. This incidentally demonstrates how the finest brains of Greece would rely on myths to support their elaborate reasoning.

Prometheus is a titan, that is, technically, not a god, or at least not an Olympian. In most ancient mythologies of Indo-European people, the cosmology includes a primordial war between two races of deities, the winning side becoming the target of worship for humans; scholars have traced this theme from Hinduism (Devas vs Asuras) to Norse mythology (Gods vs Giants), including even Christianity, with Lucifer and his demons on the losing side. In ancient Greece, titans are descendants of Gaia and Ouranos, and led by Cronus, who rules from mount Othrys. Cronus is described as a paranoid tyrant with the incongruous habit of swallowing his children; but his wife Rhea tricked him to save their youngest son Zeus, succeeding in substituting rocks for the infant god. Zeus was raised in secret, then gathered a party of upstart rebels on mount Olympos, and launched a war against the titans. That war is called the *Titanomachy* (which simply means "war of the titans") and lasted for ten years, culminating in the defeat of Cronus, and release of Zeus's siblings from Cronus's stomach (being digested for decades is only a temporary setback for a deity).

Among titans were four brothers, sons of Iapetus and Clymene (or Themis, depending on the author who relates the story):

---

[2]"Sophist" became a bad word, synonymous of one who indulges in flawed logic for dubious rhetorical goals; but that was much later on.

Figure 1.1: Plato (about 428 BC – 348 BC). This bust is kept
in the Pio-Clementino museum in Vatican; it is a Roman copy
of a Greek original from the last quarter of the 4th century BC
– it thus dates from after the death of Plato, but close enough,
so that the sculptor may have met Plato in person. That copy
has been (wrongly) inscribed much later on with the name of
Zeno, another philosopher.

Menoetius, Atlas, Prometheus and Epimetheus. During the Titanomachy, Menoetius is killed by Zeus, then exiled to the Tartarus along with all other titans (again, death is a mere transitory inconvenience for an immortal being). Atlas gets an unusual fate: instead of joining the Tartarus, which is a grim place of torment and suffering, similar to the christian Hell, Atlas is tasked with supporting the sky on his shoulders. According to myths, he still does; he was relieved from that duty by Heracles for only a few moments[3]. Prometheus, however, was smart; he saw how the war was going, and deftly switched sides. Since he ended up allied with the winners, he avoided being cast to the Tartarus, and was allowed to hang out with the Olympians afterwards, an arrangement very similar to the status of Loki with regards to Norse gods. Prometheus trailed his brother Epimetheus with him.

Prometheus is thus a character with a dubious pedigree; he is a turncoat, and a trickster. He subsequently plays the role of a kind of counterforce to Zeus's hegemony, using his wits to thwart the pure might of the chief of all (remaining) gods. We find Prometheus again in the "Trick at Mekone", again related by Hesiod. In that story, the gods are trying to work out, when Men sacrifice animals to them, what part of the offering can be retained by the mortals for their own consumption. Prometheus finds it fit to trick Zeus: he slays an ox, then he puts all the good meaty parts in the stomach of the beast, thus disguising it as an unappetising mess of viscera. Conversely, from the bones and fat of the ox, Prometheus skillfully assembles a shining dish that makes Zeus salivate[4]. Succumbing to

---

[3]Carrying the whole sky was exactly the kind of showy feat that Heracles was very fond of; but accepting a long-standing responsibility was definitely not Heracles's style.

[4]In that episode, Prometheus applies all the recipes of modern fast food restaurants.

the false promise of taste of the bone-and-fat concoction, Zeus leaves the meat to the mortal worshippers. This established a precedent, and, indeed, in ancient Greece, when cattle was offered to the gods, the priest would burn the fat and bones, letting the smoke convey their share to the deities; the leftovers, i.e. the meat, would then be ritually eaten by the participants. That was a very convenient arrangement, from the point of view of Men. Zeus was not pleased.

We then arrive at the myth reported by Plato in the *Protagoras*. Having won the war against the titans and established their hegemony, the Olympians set out to creating mortal creatures, i.e. animals and Mankind. Zeus was a true Lord: he rarely did any hard work himself. Instead, he delegated. He thus entrusted Prometheus and Epimetheus with actually doing the job of populating the Earth with all manners of animate beings, with their proper attributes. To Greek ears, "Prometheus" and "Epimetheus" are very meaningful names: they mean "fore-thought" and "afterthought", respectively[5]. Thus, Prometheus is the smart one, and Epimetheus is a dimwit. It is then no wonder that Prometheus decided to "supervise" the operation, leaving the tedious task to Epimetheus[6].

Epimetheus set to work enthusiastically, and did so by creating *archetypes*. In the myth, the gods made a number of small clay

---

[5] There is lively debate in some specific academic circles about the true etymology of "Prometheus"; according to some theories, it would be a derivative from an old Proto-Indo-European root for "thief"; for others, it would be cognate with the Sanskrit name for a tool to light a fire. However, what matters for the history of ideas, in our case, is how the names were understood by Greek philosophers, not their original etymology.

[6] As Plato relates the myth, Epimetheus seems to be the one suggesting this repartition of roles, but Prometheus did nothing to dissuade him, and it is no stretch of imagination to suspect that the trickster titan first planted the idea in his gullible brother's mind.

sculptures, each representing an animal[7]; Epimetheus's task was to imbue each sculpture with the specific attributes allotted to that species. For instance, there was a figurine of a duck, that stood for all ducks. Archetypes are the fundamental reality of Plato's "Theory of Forms" that he uses in his works: tangible objects and ideas really are conceived to be mere projections or shadows of an entity called a Form. Every time you see a duck, you are only perceiving some aspect of the ideal Duck. That theory is meant by Plato to explain why the tangible world appears to vary: you can see small and big ducks, the ducks move and can change shape, indeed a duck can even be killed, plucked, cooked and eaten; but to Plato, these are all mere changes in your point of view. As you digest the duck breast, you are just observing from a different and varying angle the fundamental, unchanging duckness for which all ducks are proxies. That central point of existence for all that is duckish is Epimetheus's figurine.

This whole notion of archetypes, and how it influenced and still influences our perception of the world, is the subject of this book. We will be following the traces of Epimetheus's creations throughout our collective minds, and see that his legacy is in no way negligible next to that of his more flamboyant brother.

Meanwhile, Prometheus was less impressed. Indeed, Epimetheus had been happily distributing attributes to all his figurines; Plato explains how all these attributes balanced in harmony:

> *There were some to whom he gave strength without swiftness, while he equipped the weaker with swiftness; some he armed, and others he left unarmed; and devised for the latter some other*

---

[7] In Plato's terms, they were "fashioned out of earth and fire", a nifty way to describe pottery.

means of preservation, making some large, and having their size as a protection, and others small, whose nature was to fly in the air or burrow in the ground; this was to be their way of escape. Thus did he compensate them with the view of preventing any race from becoming extinct. And when he had provided against their destruction by one another, he contrived also a means of protecting them against the seasons of heaven; clothing them with close hair and thick skins sufficient to defend them against the winter cold and able to resist the summer heat, so that they might have a natural bed of their own when they wanted to rest; also he furnished them with hoofs and hair and hard and callous skins under their feet. Then he gave them varieties of food-herb of the soil to some, to others fruits of trees, and to others roots, and to some again he gave other animals as food. And some he made to have few young ones, while those who were their prey were very prolific; and in this manner the race was preserved.

(This excerpt is from the classic translation of the *Protagoras* by Benjamin Jowett.)

Interestingly, this paragraph shows that even in the days of Plato, in the 4th century BC, the concept of species extinction was already known and reasoned about[8]. We'll get back to that in a later chapter.

While Epimetheus did an apparently good job, he was working with a finite set of attributes to distribute, and, in his careless-

---

[8]Jowett uses the term "race" because that was still a loose synonym for "species" in the 19th century.

ness, he ran out before the end of the task. When Epimetheus had to equip the last figurine, which was that of Man, his bag of attributes was empty. Man was some hairless ape, weak, unprotected from elements, with no innate ability at hunting preys or evading predators. If left in this state, Man's prospects would be grim indeed. Prometheus knew that the gods would not be happy with that outcome. He thus felt compelled to improvise: in order to give Man a fair chance at survival, he gave him the mechanical arts and the fire, both stolen from Hephaestus. The myth is not very clear about whether the arts and the fire were two distinct things, or two aspects or metaphors of a single attribute now best known as "intelligence"[9]. In any case, this was not part of the attributes that the two titans were supposed to grant. Prometheus had tried to cover his brother's blunder with what turned out to be an even bigger blunder. Zeus, again, was not pleased.

From a human point of view, Prometheus is to be praised, since his theft allowed Mankind, alone among all animals, to obtain a small part of divinity. This part of the myth sets Men apart – we'll get back to that later on. Gods saw this differently: to them, this was pure thievery. Zeus thus punished Prometheus in an uncommonly rash way: Prometheus was chained to Mount Kazbek, whereupon an eagle would come daily to devour Prometheus's liver; owing to the immortal nature of the titan, this essential organ would simply regrow overnight, thus enabling Prometheus to suffer again the next day. What the eagle thought of this arrangement is not recorded, but it apparently strived on this liver diet until Heracles came by

---

[9]As Plato relates it, this does *not* include political wisdom, which was exclusive to Zeus himself.

(many years afterwards) to slay the bird and free Prometheus[10]. Interestingly, while Prometheus was thus despatched, Epimetheus emerged unscathed; apparently, his lack of wit was notorious, and punishing him would have been the divine equivalent of kicking a spaniel.

The God of Jews, Christians and Muslims is the epitome of Justice. Zeus was not. Though he was supposed to wield the archetypal "political wisdom", he could be quite nasty and unfair at times. After having dealt with Prometheus, Zeus found it fit to extend the retaliation to Men as well, though they were hardly at fault in the whole story. Zeus thus made Hephaestus create Pandora, the first woman; the other gods gave her some specific attributes, such as a seductive appearance and a deceitful mind. Zeus finally equipped Pandora with a jar (later mistranslated as a "box") containing "all the evils" that would plague Mankind. To send Pandora upon Mankind, Zeus employed that useful idiot, Epimetheus: he offered her to him as a bride. Despite having been warned by his brother never to accept anything from Zeus, Epimetheus took Pandora, whereupon she promptly opened her jar, leaving all the evils out, except Hope, that remained stuck to the bottom. From that point onward, Men would have to toil to extract their subsistence from the Earth; they would become sick, and die.

The myth of Pandora and her box led to a huge amount of commentaries. It is of course literally dripping with misogyny: Pandora is treacherous and inconstant, and brings woe through her uncontrolled curiosity. Some versions of the myth even suggest that Pandora and her box are the same entity, thus marking Women as the scourge of Men. The exact nature of the "evils" is

---

[10] Possibly. In other accounts, Heracles felt content with killing the eagle and leaving Prometheus chained on top of a tall mountain. Heracles was, after all, Zeus's son.

not disclosed; neither is it said what Hope was doing among the evils (some scholars suspect a mistranslation of a more ominous term). We still see Epimetheus as the dope through which deities enact their plans. This went further: as Epimetheus's wife, Pandora gave him a daughter, Pyrrha, who married Deucalion, son of Prometheus. The two cousins then were the only survivors of the Greek version of the Flood, and thus became the ancestors of all humans who thereafter repopulated the Earth.

The whole narrative is somewhat contradictory: while he was chained and used as a self-replenishing bird feeder, Prometheus could hardly have been concocting meat pies and fat decoys; thus, the trick at Mekone must have occurred before the theft of fire. However, how could the gods debate about the proper way for Men to make offerings, before Men were granted intelligence and the fire, both necessary components for the whole concept of making an offering to the gods? But remember what I wrote above: story inconsistencies were not considered in ancient Greece to be detrimental to the truth value of the myths.

Among this set of myths, we shall retain a few salient points. The first one is that while Zeus is top god, he is not omniscient and does not even maintain good control of things. In Homer's Iliad, there is a passage where Zeus threatens the other gods:

> *"Let none of you neither goddess nor god try to cross me, but obey me every one of you that I may bring this matter to an end. If I see anyone acting apart and helping either Trojans or Danaans, he shall be beaten inordinately ere he come back again to Olympus; or I will hurl him down into dark Tartarus far into the deepest pit under the earth, where the gates are iron and the floor bronze, as far be-*

neath Hades as heaven is high above the earth, that
you may learn how much the mightiest I am among
you. Try me and find out for yourselves. Hangs
me a golden chain from heaven, and lay hold of it
all of you, gods and goddesses together – tug as you
will, you will not drag Jove the supreme counsellor
from heaven to earth; but were I to pull at it myself
I should draw you up with earth and sea into the
bargain, then would I bind the chain about some
pinnacle of Olympus and leave you all dangling in
the mid firmament. So far am I above all others
either of gods or men."

(This excerpt is from the beginning of book VIII of the Iliad, as translated by Samuel Butler. Butler uses the Latin names, thus Zeus is called Jove.)

So much for "political wisdom". Zeus is deploying arrogance and threats of physical abuse to assert his alpha male status; he is more a schoolyard bully than the wisest of monarchs. The myths of Prometheus clearly demonstrate that Zeus, for all his strength, can make mistakes and be outsmarted. Prometheus and Epimetheus together summarise the relationship between Men and the gods: like Epimetheus, mortal Men shall cower and grovel before the gods, and fear their might; like Prometheus, they should also strive to outthink the gods and improve their condition by using their intelligence in unforeseen ways. When Men interact with Olympians, they do so mostly contractually. Gods do not see what lies in the minds of Men, or they do not care. What matters to gods is the outward show of respect, and they grant boons based on the size of offerings. It is up to Men to work under these conditions, and not incur Zeus's wrath.

To sum it up, while Zeus is the mightiest god, he is in no way perfect; his creations can be flawed, and often are. You just have to exercise a lot of caution before telling it plainly, because Zeus can be a bit sanguine. For some reason, poets alone can get away with explaining in full graphic details how Zeus can be, at times, dead wrong.

The second point is that while all animals were neatly made into archetypes (Plato's Forms), the distribution was overlooked by an intellectually challenged deity. Not only is Zeus imperfect and prone to blunders, but he delegated the task to an inept assistant who is not even a true god, but a reformed titan. It is thus unsurprising that the end result may be somewhat unsatisfactory. A few centuries later, Christians would have a lot of intellectual trouble with the "Problem of Evil", i.e. how a perfect benevolent God could allow evil to exist at all; but in ancient Greece, evil is an understandable outcome given the high level of incompetence in the divine top management. Pandora's box even seems redundant.

Plato died around 348 BC, in a world that was about to be politically overwhelmed, first by Alexander's conquest of all of Greece, then Persia and beyond, triggering vast exchanges of ideas between Greece and India, where Buddhism was being developed. The resulting greco-buddhist culture had long-standing influences in art and philosophy throughout near and middle East. Alexander's empire was short-lived, and after his death, Greece turned back to the Aegean and Mediterranean seas, where new powers were trying to establish their hegemony. After a couple of centuries, an upstart town called Rome succeeded in building a much sturdier imperium that lasted for more than half a millennium.

Romans were very good at administration and bureaucracy, but they felt somewhat culturally subservient to Greece. They absorbed their art, their philosophy, and their gods; they understood that their own chief deity, the sky god Jove, was really the same dude as Zeus. This resulted in a much enlarged mythology, since "all stories are true". While the once free Greek cities had become mere provincial towns in the Roman Empire, stripped of their former political power and almost reduced to quaint places to indulge into artistic delight, the Greek ways of thinking, their philosophy, had conquered the minds of the Empire that, at that time, was comprising about a quarter of the total Earth population.

This held for a few centuries, until the rise of Christianism.

✌ 2 ✌

The Perfection of Creation

In the last decades of the first century BC, the Roman Republic had established its power over all areas around the Mediterranean. It then went through Caesar's attempt at dynastic preeminence and subsequent assassination; civil war ensued; Caesar's great nephew and adopted son Octavius restored order and became the first Roman Emperor, under the name of Augustus. Augustus was very successful at increasing the integrity of the Empire, tightening integration of dependent territories.

One of the most troublesome territories was Judea. It was populous: it might have contained up to 10% of the population of the whole Empire. It was rich: its position between the old civilisation areas of Egypt, Mesopotamia, Assyria and Anatolia ensured for several millennia that a lot of commerce would go through it[1]. Most inhabitants of Judea, the Jews,

---

[1] This strategic position also ensured that it would be perennially disputed territory, and, indeed, Judea enjoyed true political autonomy only when all great powers in its vicinity would face deep internal trouble, as it happened at the end of the second millennium BC; Judea became the Kingdom ruled by Saul, then David and Solomon.

were professing a strange religion: they asserted that their god, Yahweh, was the One True God; that deity not only required exclusive worship, but his followers were supposed to deny the actual existence of any other god, including some notoriously powerful and irate deities like Zeus (known as Jupiter or Jove in Rome). As part of the settlement that turned Judea into a client kingdom for Rome in 63 BC, Judeans had negotiated an arrangement by which they could keep on worshipping only their jealous god. This was keeping Jews out of the Empire's mainstream spirituality, and helped them maintain a sense of self-existence as a specific entity within the Empire. From the Roman point of view, this arrangement was tenable because most jewish sects were non-proselyte, so there was little risk of contamination of such "national" ideas[2]. Also, while Romans followed the same gods as Greeks did before them, they were also a bit more down-to-the-earth and were willing to incur a bit of Jovian displeasure if it could keep the wealth of Judea within the Empire.

Still, Judea suffered from chronic unrest, which at times erupted into full-fledged revolt. The first of such rebellions began in 66 AD, and was allowed to linger on for a few years because Rome was going through a dynastic change: after the death of Nero, the last of the Julio-Claudian emperors, three short-lived emperors took the office successively, before it was claimed by a strong-willed, heavy-handed Vespasian, who restored order. Vespasian had been previously involved in trying to contain the jewish rebels. Once in power, in 69 AD, Vespasian sent his very capable eldest son Titus to finish quelling the uprising. Titus did so, and, very significantly, burned down the Temple. The

---

[2]Speaking of "nations" at that era is an horrible anachronism, but it still conveys the idea that Rome was hard at work tracking seditious behaviours that would compromise the Empire's integrity.

Temple, rebuilt by Herod where Solomon's Temple was said to have stood, was the geographical centre of the jewish spirituality, and a symbol of their special status within the Empire.

This destruction of the Temple had far reaching consequences. Most importantly, it forced the jewish religious authorities to find another spiritual centre for their faith, and they opted for a non-tangible one: thereafter, their faith would be based on orthodoxy. Before Titus's bonfire, a Jew was whoever performed his yearly worshipping at the Temple; afterwards, "true" Jews would be those who keep to the letter of Mosaic law. Thus, a large number of peripheral, heterodox sects were kicked out and forced to find some spiritual anchor elsewhere. This included a small group who were beginning to be known as "Christians". These were the former followers of a reformist rabbi known as Jesus, who ran afoul of the jewish religious authorities of that time, and was sentenced to die the death of the lowest of criminals: the crucifixion. Four decades afterwards, his pupils were still propagating his teachings, but they were divided: some thought of themselves as completely jewish, and understood Jesus's message as being part of the continued story of the Jews; others were proselytes and claimed that non-Jews could and should partake to the universality of the new doctrine. Titus's rashness indirectly but decisively ended the debate: since Jews were reorganizing their spirituality around strict adherence to the law of Moses, there was no room for heretics who believed that an ignominiously executed criminal could be the most expected Messiah. Christianism thereafter had to become a determinedly universal endeavour and propagate through the "gentiles", that is the polytheistic, pagan masses of the Empire. The proselyte Christians, led by the example of Paul of Tarsus[3], set out to convert the Roman world to Christianism.

And so they did. But it took time: the real breakthrough came at the beginning of the fourth century AD, when Emperor Constantine declared himself a Christian. At that time, an estimated 15% of the Empire's population was christian. Less than a century later, in 393 AD, the figure would have reached more than 80%, and Emperor Theodosius made paganism mostly illegal, thereby marking Christianism's victory.

Thus, the Greek world turned christian. What did it imply for how the people envisioned living creatures? What became of the notion of archetypes?

Christianism, being an offspring of Judaism, inherited its mythological corpus. Contrary to Greek paganism, christian mythology is organized and curated: religious authorities are keepers of myths, and they decide, through much thinking, talking, assessing, and presumed divine inspiration, which myths are true and which are not. This is a marked departure from the chaotic production of self-appointed poets, and it goes hand in hand with monotheism: if gods could be fake, so could myths. Monotheism is, at its core, the forceful insertion of hard, physical logic into the spiritual world.

Another factor greatly contributed to make Christianism a very intellectualistic religion. As we saw above, after the destruction of Herod's Temple, Christians turned to all-out proselytism, which entails convincing other people. At that time, refined Greek philosophy, combined with the definitely bureaucratic Roman attitude, made the elites of the Empire give high value to logic and rational thinking. Proselytism, at least in the

---

[3]He was already dead at that time, but his fierceness in his proselytism was famous.

upper sphere, had to be done with robust argumentation. And so it happened: early christian authors like Origen and Tertullian engaged into heated but highly rhetoric debates with champions of paganism, for example Celsus and Porphyry[4].

For such reasons, Christianism, as it gained hold throughout the Greek and Roman society, was also becoming a highly structured and gigantic intellectual construction. One of the momentous decisions of Constantine, once comfortably in power after the defeat of the last of his co-Emperors (Lucinius, in 324 AD), was to decree and organize the First Council of Nicaea. In 325 AD, bishops from the whole Empire met in the town of Nicaea, with the goal of establishing a definitive, unified doctrine, in particular with regards to the nature of Jesus, the proper computation of the date of Easter[5], and the first draft of the canon law.

The Council did not apparently discuss the establishment of a formal list of canonical books for the Bible, but the process was already ongoing and well advanced at that time, and Constantine was again applying pressure for the production of such a list. Indeed, a large number of texts had appeared in the second century AD and later, in particular gnostic writings. Constantine wanted a strong, unified Church, and such a goal was not compatible with an haphazard proliferation of conflicting doctrinal variants. For Christians this was all heresy, to be fought against. The process by which the Bible canon was defined really was a patient but thorough removal of all works that were

---

[4]Due to the ultimate political victory of Christians, almost none of the anti-christian works survived. We know of Celsus only from the excerpts quoted by Origen in his *Contra Celsum*.

[5]Literally: the *computus* was in medieval times the calculation of the date of Easter; the modern term "computer" is derived from that name. The Council of Nicaea did not produce complete rules for the *computus*, but at least it initiated the effort.

suspected of deviating from the original teaching of Jesus, or not part of what Jesus, as a jewish rabbi, would have accepted as Holy Scripture. In practice, nothing that was written later than the first century AD made it to the definitive christian Bible[6].

Thus, after a few centuries, the nascent Church acquired a consensual, canonical list of Holy Scriptures that sets the accepted mythology. Let's see what that mythology says about the concepts covered by the Greek myths about Prometheus and Epimetheus.

It all begins with the start of the very first book, Genesis[7]. Genesis begins with a relation of the Creation, and, confusingly, relates it again from a different angle. For the Bible excerpts below, I use the King James Bible, commissioned by King James I of England and Ireland (also known as King James VI of Scotland) in 1604 and completed in 1611.

In the first Creation tale, God first creates heaven and earth. The earth was *without form*, and dark, and wet too, since "the Spirit of God moved upon the face of the waters". Then God creates light, and He does so by *saying*: to speak the name of a thing is sufficient to make it pop into existence. Once there is light, there are days and nights, and thus begins the counting of time. During the first six days, God makes all the work. At day three, God makes dry land and plants. During day five, God fills the sea with marine animals, and the sky with birds.

---

[6] Of course there is no such thing as a single definitive christian Bible. The Roman Catholic Church has one, that slightly differs from that of Orthodox Church and other Eastern Churches; and a myriad of Protestant faiths have made almost countless variants. All these lists still agree about most books.

[7] The first book in canonical order, that is. According to scholars, while the Genesis relates the creation of the world and appears to be a very ancient base story, the text itself has probably been put in written form in the 6th century BC, at least two centuries after most of the Psalms and some other books such as Amos and Isaiah.

The next day, God finally produces land creatures, and Man. In that first version, Man is created as "male and female", i.e. men and women are both made at the same time. God finally puts Mankind in charge:

> *And God blessed them, and God said unto them, Be fruitful, and multiply, and replenish the earth, and subdue it: and have dominion over the fish of the sea, and over the fowl of the air, and over every living thing that moveth upon the earth.*

And then He goes to rest. On the seventh day, God invents the week-end, dedicated to everything but work.

Then begins (chapter 2, verse 4) the second telling of the Creation. This time, the events are not assigned to specific "days", and they are told in an order that is not fully compatible with the first tale[8]. This time, God creates mostly plants, and in particular a bountiful garden in a place called Eden. God then creates Man (the one male called Adam, this time) and puts him in the garden. Then, to help man and keep him occupied, God proceeds to create all land beasts and all birds (there is no mention of sea animals in this second tale). Like Man, the beasts and birds are formed out of "the dust of the ground" and God breathes life through their nostrils; but Man is involved in the process:

> *And out of the ground the LORD God formed every beast of the field, and every fowl of the air; and brought them unto Adam to see what he would call them: and whatsoever Adam called every living creature, that was the name thereof.*

---

[8]Official Church explanation is that the "days" are symbolic, not chronological.

However, among all this duly named bestiary, there was no appropriate help meet for Adam; so God anaesthetised Adam and surgically extracted one of his ribs, out of which God made the first woman, as yet unnamed. Adam wakes up, and, probably out of habit (he had just named all the animals of the world), proceeded to name the woman, maybe unimaginatively, "Woman", because "she was taken out of Man". Adam will name her again in chapter 3, verse 20, this time calling her "Eve"; we are to understand that in that second instance, Adam names her as an individual, not as a broad category.

Figure 2.1: Adam in the garden of Eden, before things go sour. This scene was etched by Johann Elias Ridinger circa 1750; it depicts the assumed harmony of all species created by God.

This double tale of Creation has many similarities, as well as striking differences, with the Epimethean myths described in the previous chapter. Archetypes are again made into existence: in the first Genesis tale, all animals and plants are created by species; in the words of the King James Bible, "God created [...] every winged fowl *after his kind*". This "after his kind" expression is used for all living creatures. Animals of the land and the sky are built out of dust, which recalls the clay figurines that were made by the gods and given to Epimetheus. In the Genesis, Adam plays the role of the titan, as he "names" the various species. As we saw, at the time the Genesis was written, naming a thing was very much like creating it; that's how God himself proceeds. This naming was thus quite akin to the distribution of attributes to animals by Epimetheus. However, the change in terminology was the prelude to a weakening of the role of Adam in later interpretations: centuries later, Adam would be understood as having merely put labels on God's creatures.

In the Genesis, Woman is, as in the Greek myth, given to Man. However, this is not done as part of a godly vengeance; in fact, the Woman is the missing half without which Man is not complete. The old idea that women are bad news still appears in Genesis, but later on: the first woman is tricked by the Serpent into transgressing God's orders about a special tree planted in the garden of Eden, and she then "contaminated" her husband. In chapter 3, verses 11 and 12, we see that Adam is quite adept at shifting blame:

> *And he said, Who told thee that thou wast naked? Hast thou eaten of the tree, whereof I commanded thee that thou shouldest not eat?*
>
> *And the man said, The woman whom thou gavest to be with me, she gave me of the tree, and I did eat.*

As we see, there are many similarities between the Greek and christian stories about Creation. From a religious point of view, this is expected: all of Mankind lives in the same world, and the believers of other creeds, while sorely misguided, may still have remembered a few correct facts about the One True God (or set of gods, where applicable). In a non-religious analysis, these similar mythologies will be said to derive from older common ancestors, and, indeed, mythologies, like languages, are a useful tool to work out the broad lines of population moves and settlement areas in ancient times.

Yet there are also great differences between the christian and the ancient Greek tales. The most important of these is that in the christian story, God, who is omniscient, just and loving, is doing most of the work. In the Greek myth, the gods are a bunch of revellers who set out to serious work only when they have no choice; Zeus is depicted as very powerful, but not wise beyond measure. In any case, Zeus delegates to a pair of titans, i.e. second-class deities, and the one who ends up doing the job is notoriously stupid. He indeed botches things up, and his brother tries to cover him with an extra blunder. The whole ordeal reeks of mismanagement, to use nice words. Under such auspices, it is understandable that the end result may leave to be desired. This is totally different in the jewish and christian version, where the Creator is, by definition, incapable of doing anything that is not end-to-end perfect. All animal species were designed by God himself, so of course they are exactly what they should be. This specific point will plague scientists later on, when they begin to encounter the concept of extinction.

Another notable difference with the Greek myth is that Man is explicitly given dominion over animals and some plants:

*And God blessed them, and God said unto them,
Be fruitful, and multiply, and replenish the earth,
and subdue it: and have dominion over the fish of
the sea, and over the fowl of the air, and over every
living thing that moveth upon the earth.*

*And God said, Behold, I have given you every herb
bearing seed, which is upon the face of all the earth,
and every tree, in the which is the fruit of a tree
yielding seed; to you it shall be for meat.*

*And to every beast of the earth, and to every fowl
of the air, and to every thing that creepeth upon the
earth, wherein there is life, I have given every green
herb for meat: and it was so.*

This gives explicit permission to Man to eat fish, birds and anything that moves on the ground, and also plants that produce
seeds. Animals, on their part, are supposed to eat grass. The
Lord did not specify what would happen if Man or some animal
took it upon himself (or itself) to eat something else; weirdly, the
case of carnivorous animals is not covered[9].

In the Greek version of the Creation, Man is far from being the
dominant species. In fact, thanks to Epimetheus's poor planning, Man starts at the very bottom of the food chain, with ab-
solutely nothing to help him survive. Prometheus stole intelli-

---

[9]The subject of what is ritually acceptable food for believers is a large subject and some rules, in particular jewish rules about what is Kosher and what
is not, are bewildering in their complexity. In biblical terms, such restrictions
are part of the "covenant" between God and his chosen people, and several
such covenants are made throughout biblical times. The last one was established by Jesus, and it explicitly allows all food items, while Jews, who do not
recognise Jesus as the son of God, operate under the previous versions, in
particular Mosaic law, that lists a number of animals as ritually impure.

gence from the gods and gave it to mortal men, thereby allow-
ing them to live, but they were still, in Plato's terms, "destroyed
by the wild beasts, for they were utterly weak in comparison of
them". Thus, dominion over the living world was slow to come,
and, in any case, originated from thievery. This was not the will
of the gods. This contrasts with the christian outlook, whose
mastery over all of Nature is taken as granted, that is, explicitly
granted by God himself.

In the Greek mythology, Zeus was angered by the hubris of the
"Pelasgians", a catch-all term for designating populations who
inhabited the Aegean region before the Greeks. Always a master at subtlety, Zeus decided to kill them all by drowning, and
made rain pour so much that a great flood ensued. However,
Deucalion was warned by his father Prometheus[10] and he built
a chest, in which he floated with his wife Pyrrha until Zeus's
anger, and the waters, receded. One more time had Prometheus
succeeded in foiling Zeus's plans. But Deucalion and Pyrrha
smartly thanked Zeus, not Prometheus, for sparing them, and
this mollified enough the leader of the gods that he let them live
and even showed them how to spawn mankind in a more expe-
ditious way than the usual biological arrangement[11].

The biblical flood myth is again similar, with some modifications. In the Bible (Genesis, chapters 6 to 8), God was angered at
mankind for their wickedness[12] and decided to start again with a
clean slate. But God noticed that there was an uncorrupted man

---

[10]Since Prometheus was supposedly chained and sustaining daily liver
damage at that time, one must assume that Deucalion was a good son and
visited his father at least occasionally.

[11]Deucalion and Pyrrha just threw stones over their shoulders, and the
stones became new fully grown men and women. Very convenient.

[12]While *hubris* is the capital sin for ancient Greeks, it is less so for Jews and
Christians; the Bible does not give details about the wickedness that deserved
such a radical treatment.

called Noah, and he decided to spare him. God himself warned Noah and instructed him to build the famous Ark. The Ark was to be vastly enlarged with regards to Deucalion's chest; its length was to be three hundred cubits, and its breadth fifty cubits, and it should contain three stories of ten cubits in height each. The biblical cubit is estimated at about eighteen inches, so the total living space in the Ark would amount to about eighty thousand square feet (about 7,400 square metres) if we assume a roughly elliptic shape for the Ark. This is roomy for Noah and his immediate family (his wife, his three sons, and their wives), but God had other plans: Noah was to save a pair of each of the living animal species (excluding sea creatures, who did not suffer from the Flood). This is described in Genesis, chapter 6, verses 19 and 20:

> *And of every living thing of all flesh, two of every sort shalt thou bring into the ark, to keep them alive with thee; they shall be male and female.*
>
> *Of fowls after their kind, and of cattle after their kind, of every creeping thing of the earth after his kind, two of every sort shall come unto thee, to keep them alive.*

with the extra provision (in chapter 7, verse 2) that for "clean" animals, i.e. those which are ritually pure, there will be seven pairs of each[13]. Now the Ark begins to look quite cramped, especially if you consider that elephants and giraffes are technically supposed to be part of the salvage operation. It all went well, though. God made rain pour for forty days and night, and it took a hefty one hundred and fifty days for the waters to abate

---

[13]This specific provision might be interpreted as a veiled directive for protection against inbreeding when raising cattle.

and dry land to appear again, at which point Noah could disembark and let his menagerie loose, then set out to replenish the Earth.

Figure 2.2: The animals entering Noah's Ark, depicted by Jan Brueghel "the younger" in the 17th century. The Ark itself is barely visible in the background. Brueghel made a spirited effort at symbolizing all animal species, including a pair of red aras, parrots indigenous to tropical America that were unknown to the Old World in the ancient "biblical" times. One may also note other exotic animals such as camels, elephants, ostriches, and even some bats.

The biblical flood myth is remarkable in that it indirectly establishes a founding ancestor structure for all animal species, not just mankind. In chapter 7, the chosen pairs come by themselves; Noah does not have to round them up. It follows that God chose the pair, for each species, that would engender after the flood all the new members of that species. It is as if each individual archetype got a fresh layer of divine paint. We also notice that, once again, God maintains leadership on events, contrary to Zeus: Noah is saved because God decided it so, and with considerable planning. As a minor difference, Zeus's flood was finished within nine days (so it says, at least, in Ovid's *Metamorphoses*), but God's business takes a lot more time.

From that point onward, the relationship between Mankind and Nature, in the christian world, was set: living creatures were ordained into distinct, well-defined species by God in person, and He even confirmed them in Noah's Ark. At no point was the Eternal outsmarted or outmaneuvered by anybody. The result can thus only be perfectly harmonious. If anything went wrong, it could only happen through Man, who, as a special gift, had free will, which included the ability to be tempted by the usual troublemaker known as Satan.

❧ 3 ❧

Raiders of the Lost Philosopher

"Empires wax and wane", starts the 14th century Chinese novel *The Romance of the Three Kingdoms*[1]. This rule also applies to the Roman Empire. Let's see, in broad terms, how it went.

The Roman Empire started in Rome, in the region of Italy known as Latium. At the end of the first century BC, Rome conquered all areas around the Mediterranean sea; this included half of modern Europe, up to the Rhine and Danube rivers, so large areas like the Gauls (today's France, Belgium and part of Netherlands) and Iberia (Portugal, Spain) were under Roman rule. Most of the Greek world was also part of it: what we now call Albania, Greece, Serbia, Bulgaria, Turkey, Syria, Lebanon, Israel, Egypt and Lybia were all incorporated in the Roman polity. The conquests of Alexander the Great, in the fourth century BC, had extended the Greek cultural area very far to the East, where it encountered Buddhism; however, Alexander's Empire was very short-lived, since it broke up as soon as Alexander died.

---

[1]At least in the 1959 translation by C.H. Brewitt-Taylor. Other translations may disagree.

So the Eastern half of the Empire was "Greek". Of course, its inhabitants spoke many different languages and dialects; but if you were rich and educated, and lived in the Eastern half, and wanted to talk about art or philosophy, then you did so in Greek, and it was Greek art or Greek philosophy. Conversely, in the Western half, which comprised a larger but less populous and less urbanised area, Rome itself was the cultural reference, and the language for intellectual activities was Latin. The boundary between Latin and Greek areas was roughly North-South and went between the modern countries of Croatia and Serbia, and right through today's Bosnia.

For its first four centuries of imperial existence (from Augustus to, say, Valens), Rome's worst enemy was Rome. Technically, Rome ran several campaigns against its neighbours; the strongest of them was the Parthian and Sassanid Empire, that battled against Rome for centuries over the control of Mesopotamia (modern Iraq). After the death of Alexander, his generals split up the conquered territories; Seleucus obtained Babylonia, and pursued an aggressive policy of conquest, resulting in an Empire that would somehow rebuild the Achaemenid Persia that Alexander had defeated. Later, the Seleucids' power waned because of belligerent neighbours (including Romans) and dynastic wars, and by 100 BC the descendants of Seleucus controlled only part of modern Syria. The lost territories passed under the control of the Parthians. Parthia was a region located in what is now northern Iran; in 247 BC, at a time when the Seleucids were in disarray (their capital had just been seized by Egyptians), Parthia declared its independence, just to be conquered a few years later by Parni nomads. Arsaces, leader of the Parni, then used Parthia as his political basis to build yet another Empire, thereafter known as the Parthian Empire, or the Arsacid Empire. Parthians ruled Persia for more than four centuries, and regularly fought against Romans. Following a well-established pattern, the Parthian Empire was destabilised by chronic rebellions and internal strife, and in 224 AD, the Arsacid dynasty was overthrown by Ardashir, a local leader from the Pars province (now known as Fars, in south-central Iran). The new dynasty, the Sassanids, was named after Ardashir's grandfather Sasan; it pursued the policy of opposition to the Romans, until the islamic conquest in the seventh century.

However, despite the Persian efforts at war, when a Roman legionary was fighting someone, it was more often than not another Roman legionary, who had pledged himself to another candidate for the supreme office. How often internal strife happened depended on the control of the Emperor over the legions and the administrative apparatus of the Empire. The Nerva-Antonine dynasty, covering almost all the second century AD, was especially successful at maintaining internal peace and cohesion. The third century AD was much unlike the previous; from the death of Severus Alexander in 235 AD to the advent of Diocletian in 284 AD, about twenty-three incumbents of the imperial throne ruled along a pattern that can only be described as nasty, brutish and short: some military commander is proclaimed Emperor by his troops; the usurper then goes to battle against the current Emperor; when he takes the office, he must run throughout the Empire to quell uprisings, until one of the usurpers succeeds, either by killing him in battle, or through some cheaper assassination. Of the twenty-three or so who could at one point claim to be "the" Emperor, only three managed not to be murdered or slain in battle, and they did so only by dying faster of something else: two from an unknown plague (probably smallpox), and one, originally enough, by being struck by lightning.

In the course of the fourth century AD, Rome was again bene-
fiting from internal stability under the strong rule of Diocletian
and his successors. It was also the time when it became christian. Then, in the second half of the century, the demographic
pressure from the populations North and East of the Empire
began to steadily increase. For several reasons, including possible climate changes, population was on the rise in Northern
and North-Eastern Europe (from modern Germany to Ukraine
and Russia) and this prompted vast migrations of people who
wanted new lands to settle. These people looked upon Rome as
the centre of civilisation, and they wanted to be part of it; they
were ready to use force if necessary. Rome was trying to manage
the flow by either giving away some free lands in border areas,
or by cutting the migrants in pieces with its legions. However,
the newcomers were so numerous that it couldn't last; the battle of Adrianople (378 AD) is often seen as the symbolic turning
point because not only the "Barbarians" vanquished the Roman
legions, but Rome had to give them, as the prize for peace, lands
that were formerly property of Roman citizens. To many Romans at that time, this is the year when Rome failed to protect
its own.

The worsening situation forced the Empire to split: the area
to defend was too large to be done from a unique central
authority, so, in 395 AD, the Empire was separated into its
halves, which were, basically, the Latin and Greek areas. From
that point onward, things went decidedly downhill for the
Latin half. From a geostrategic point of view, the Western
Roman Empire had a very long boundary to defend, with
relatively few population to support a strong army. A further
problem was that whenever the central leadership was not
sufficiently strong-willed, usurpers would pop into existence
everywhere; their rebellious legions would then try to march

to Ravenna[2]. When legions are busy either taking the capital by force, or preventing other legions from doing so, nobody is guarding the boundary. The fifth century AD is a long-winded disintegration; provinces are "given autonomy" one by one, while various Germanic chieftains vie for power over the shrinking remains of the erstwhile glorious Empire, until one of them, called Odoacer, decides to finally end things by sending the Imperial insignia to the Eastern Emperor Zeno in 476 AD. At that time, the Western Empire was no more than a political fiction, and was on the verge of becoming a farce. It was high time to let it die.

The political disaggregation was a facet of deeper societal changes, notably a general de-urbanisation. The population of the city of Rome, from a high mark of 1.6 million inhabitants in the second century AD, was still more than one million strong in 400 AD, but less than 100 thousands a century later, and still fewer later on. It was even temporarily abandoned after yet another sack in 546. Of course, a number of particular reasons explain at least partly that situation; for example, a million-strong city could not be fed without long-range shipments of grain, and thus the loss of the province of Africa, taken over by Genseric's Vandals, had to imply a severe reduction in population. But the trend was general throughout the West: cities turned into villages. Societies were becoming almost exclusively agrarian, and this implied huge setbacks on literacy levels. To say things bluntly, intellectualism is the product of cities: to maintain a scholar active and keep him doing scholarly things on a full-time basis, you have to feed him and fund him

---

[2]The city of Rome was the political capital of the Empire up to the end of the third century AD, but then the power shifted to other places, Milan for the Western half. In 402 AD, Emperor Honorius moved his capital to Ravenna, which was easier to defend – a clear sign that the older strategy of containing trouble on the outside of the Empire was no longer effective.

with the centralised surplus of a busy, productive population. Without cities, no intellectual elite. After the fall of the Western Empire began a period known as the "Dark Ages". The Dark Ages are not to be confused with the similarly named era which occurred in Greece after the fall of the Mycenians at the end of the second millennium BC; however, the same name was applied for pretty much the same reason: the fall in literacy and intellectual activity implied an almost total absence of written artefacts, leaving subsequent archaeologists "in the dark".

The art of reading and writing did not disappear altogether, though, contrary to the previous Greek occurrence. But non-religious themes, in particular logic, science, engineering, and speculative philosophy, were disused, and most teachings from ancient Greece were simply lost. This winnowing in the scope of knowledge was due to a number of profound reasons. The first of them is that literacy was maintained only in monasteries, a product of Christianity[3]. While even remote and "uncivilised" places like Ireland sprouted large, organised monasteries, that served as refuge islands for knowledge and culture, the monks were, by trade and conviction, a bit biased in their choice of intellectual subjects; quite logically, they mostly talked and wrote about religious matters. Latin grammar was maintained, Aristotle's considerations on physics were not. This selective focus on salvation was reinforced by a sense of urgency that was pervasive to the early christian world. Jesus had said that "ye know neither the day nor the hour wherein the Son of man cometh" (Matthew 25:13), but everybody assumed that the End of Times was nigh. Thus, there was no time to be lost in

---

[3]Monks are not a christian original invention; Buddhists had been doing the same thing since the time of Gautama Buddha. However, in the former Roman Empire, wide-scale development of monasteries was locally an unheard-of innovation.

intellectual speculation; being ready to face the Creator at any time was of the utmost importance.

Another factor was the loss of contact with the still striving Eastern Roman Empire. Of course some trade was still occurring, as well as some political relations, including the occasional war. However, the language barrier was reinforced by a theological fracture. Jesus himself had declared to the Apostle Peter: "thou art Peter, and upon this rock I will build my church; and the gates of hell shall not prevail against it" (Matthew 16:18)[4]. This really looked like an explicit mandate, from the Son of God himself[5]. The bishop of Rome, one of the five Patriarchs of the early Church, was traditionally considered to be successor of Peter, since Peter was supposed to have died in martyrdom in Rome[6]. He thus took Jesus's injunction as his cue for claiming preeminence over the four other Patriarchs (bishops of Constantinople, Antioch, Jerusalem and Alexandria, all from the Greek area), and organise the Church with himself as the single top of the hierarchy. The four other Patriarchs disagreed. The quarrel was allowed to fester and deepen, and was complemented by successive doctrinal divergences (e.g. Rome insisted on the celibacy of priests, while in the East priests could marry – and they still can). The language, cultural, political and theological boundaries piled upon each other,

---

[4]Prior to that verse, Peter was called Simon. Jesus called him "Kepa", which is the Aramaic word for "rock", often translated in Greek as *Cephas* or *Petros*, and in Latin as *Petrus* from which comes the English *Peter*.

[5]In christian theology, Jesus is both the Son of God, and God himself at the same time. Regardless of the intricate details of trinitarian doctrine, when Jesus talks, it is official.

[6]Crucified by order of Emperor Nero, Peter reportedly asked to be tied upside-down because he did not feel worthy of dying in the exact same way as Jesus. This was quite smart as well, because an upside-down Peter would pass out from the contra-gravitational blood circulation and avoid most of the suffering normally implied by crucifixion.

culminating in the Great Schism of 1054, when the Catholic and Orthodox churches split and went separate ways. Since it is considered bad form to talk to heretics, this estranging of the two main brands (at that time) of Christianity implied a cessation of intellectual intercourse between East and West.

This state of affairs began to reverse in the eighth century, and much more in the twelfth century. Several important events occurred. First, in the seventh century, in the town of Mecca in the middle of Arabia, a merchant called Muhammad began to receive revelations from God, conveyed to him by the archangel Gabriel; Muhammad wrote, under Gabriel's dictation, the Quran. Muhammad began preaching, gathered followers, founded a new religion (Islam), and, quite expectedly, ran afoul of the local authorities, like Jesus did before him. Muhammad's response, however, was very different: instead of getting himself arrested and killed, he escaped to the town of Medina in 622 AD, where he patiently built his strength and gathered an army. After eight years, he took over Mecca by force. From that moment onward, and for about a century, Islam was spread mainly through military conquest, and it was very swift. At the death of Muhammad in 632 AD, most of Arabia was islamic. In 651 AD, the battered Sassanid Empire fell to their power. By 711 AD, all North Africa up to and including modern Morocco was under their sway, and in the following decade the Umayyad caliph conquered most of Hispania (now Spain and Portugal, except the North coast). This conquest severely held in check the ambitions of the Eastern Roman Empire, that became increasingly centred on Greece; Alexandria, Jerusalem and Antioch were now in Muslim territory, which strengthened the position of Rome within the Church.

Most important for our story, the conquest of Persia implied the absorption of an intellectual elite who had kept on copying and studying ancient Greek texts. Soon, Arabic translations of Greek classics on many subjects such as medicine, mathematics, astronomy, biology… were available throughout the Caliphate. In particular, when in the middle of the eighth century the Abbasid rebelled and seized power from the Umayyad, the latter retreated to their newest and farthest conquest, Hispania, and they brought their Persian thinkers with them. In Hispania, their political story would be one of very gradual decline, known from the christian point of view as the *Reconquista*. Since the eighth century, ancient Greek philosophy again began to percolate into the Latin world, through arabic translations retranslated by bilingual christian monks, especially in Catalonia.

Meanwhile, the Dark Ages were becoming increasingly lighter. After centuries of fragmentation and large population moves, large scale political structures were again on the ascendant. The Frank king Carolus Magnus, better known as Charlemagne, built a large Empire covering at its peak an area comprising most of modern France, Belgium, Netherlands, Germany, Switzerland and Austria, as well as half of Italy and Slovenia. This large organisation favoured a rekindling of urbanity: cities began to regrow. Charlemagne's clerics further made two crucial innovations, though they seemed minor at that time: they perfected the manufacture of parchment, and they invented lowercase letters. Parchment is made of dried and scraped animal skins, and is a lot more durable than papyrus; besides, papyrus comes from a plant which was very common in Egypt but not at all in Germany, whereas animal skins are a natural byproduct of cattle husbandry. The Caroline minuscule is a script that developed as a calligraphic standard that ultimately produced modern Latin lowercase letters. The new abundance of parchment, the standardisation of lettering, and the writing speed increase allowed by the new, simplified glyphs, concurred to promote a cultural revival known as the Carolingian Renaissance. Texts, and thus ideas, were again flowing in the West.

The Year 1000 AD came by; but the Messiah did not. To most people, it would have been very appropriate, symbolically speaking, for Jesus to bring about the End of Times at that date. But he did not. Apparently, when he said that nobody knew the day and hour of his Second Coming, he really meant it. People began to think that while the Apocalypse could occur at any time, it might be quite far in the future, and thus it made sense to dedicate some thinking time to mundane matters about the tangible, physical world. Moreover, the European economy was getting better: more and more land was deforested and dedicated to agriculture; the climate was warmer[7]; the plough was improved and productivity soared. The bishop of Rome, now known as the Pope, had asserted a stronger power and was promulgating partial bans on warfare between Christians. More food and less war meant an increase in population, and, again, growing cities.

In 1095 AD, for a variety of reasons that include the need to get the idle and growing nobility busy, as well as the worsening of travelling conditions for pilgrims going to Jerusalem, Pope Urban II preached the Crusade. The Crusade was a holy war: a war fought for religious goals with little heed paid to strategy. It was a very weird affair. The islamic conquest had brought

---

[7]We do not really know why, but between approximately 950 AD and 1250 AD, the average temperature was slightly higher in Europe; this is known as the Medieval Climate Optimum. It was not as pronounced as the warm-up of the early 21st century, but it was still significant, especially for agriculture.

Jerusalem, where the Christ's tomb was located[8], into islamic lands. Muslim leaders were more or less tolerant of Christians, but when the power was seized by the Turko-Persian Seljuks in the middle of the eleventh century, tolerance was on the decline. The christian West now had the material means of meddling with the state of things in the Near East, and so they did. While the islamic world was, on average, more advanced technologically and scientifically than the West, the Franks had the advantage in warfare thanks to their heavy mounted warriors, who would perform coordinated charges with lances, and good knowledge of poliorcetics, that is "the art of destroying cities" (i.e. siege engines). From a geostrategic point of view, the Kingdom of Jerusalem was doomed: there was no way the West European countries could bring enough supplies and reinforcement to maintain a small christian area surrounded by enemies. It still took two protracted centuries for Muslims to cease their internal wars, learn the Frank art of war, invent their own version of the holy war (the *jihad*, theorised by Nur al-Din and his nephew Salah ad-Din, better known as Saladin), and kick out the Christians. During these two centuries, ancient Greek texts translated and annotated by Arabic and Persian thinkers would again flow into the Latin West, complementing and enriching the Hispanic corpus.

Thus, in the twelfth century Europe, Greek philosophy was rediscovered. As explained above, it corresponded to a time when an intellectual elite was on the rise, and was beginning to worry about non-religious matters, such as what is now in the scope of philosophy and natural science. They found that in these respects, the Bible was a bit lacking: the working idea was

---

[8]Jesus was resurrected on the third day, so of course there was no corpse or remains to be found in Jerusalem; but his very temporary resting place was still a beacon for worshippers.

that while everything in the Bible was by definition perfectly true, there could exist some truth in other areas of which the Bible says nothing. The finest brains of that era saw that there was a great wealth of ideas to be recovered from old pagan texts, in particular in the works of Aristotle. Clerics such as Abelard, William of Ockham, and in particular Thomas Aquinas, set out to reconcile Aristotelianism and Christianity, through a methodical, rational and critical intellectual process known as scholasticism. In that endeavour they retraced the path of islamic philosophers such as Avicenna (Ibn Sina, 980-1037) and Averroes (Ibn Rushd, 1126-1198), who had previously confronted Aristotelianism with islamic faith. They succeeded: Aristotle's conceptions, notably with regards to the natural world, were adopted in Western Europe in the late Middle Ages. This is where we again pick up the trail of the idea of archetypes.

Aristotle was born in Chalcidice in 384 BC. Chalcidice is a peninsula overlooking the Aegean sea in northern Greece; by Greek standards in the early fourth century BC, it was not truly Greek, since it was closer to Macedon, then viewed as mostly foreign. When he was eighteen, Aristotle moved to Athen to join Plato's *Academy*, where he remained until Plato's death in 348 or 347 BC. Afterwards, Aristotle founded his own place of teaching called the *Lyceum*. A decade after, he formalised his thoughts as the Peripatetic school, named after the Greek word meaning "to discuss while walking"; apparently, Aristotle liked to move around while talking[9]. Aristotle died in 322 BC, but his ideas survived, especially because he wrote a lot. Previous generations of philosophers were wary of writing things down; Socrates was

---

[9]In much later times, "peripatetician" was also used as a slang term for a prostitute who seeks customers by walking around streets at night. But at the time of Aristotle, there was no such association.

Figure 3.1: Saint Thomas Aquinas, painted by Carlo Crivelli in 1476, almost two centuries after his death. Aquinas holds a miniature church and a book as a metaphor of his use of both faith and reason to reconcile Aristotelianism with Christianity.

even opposed to the concept because it encouraged laziness. But by Aristotle's time, texts were all the rage.

Aristotle worked out logic, which is what allows you not to be swindled by smooth talkers. When scholastics read about logic, it sounded like music to their ears: logic is what makes all the difference between a cohesive reasoning, and a mere haphazard juxtaposition of concepts. Armed with his logic, Aristotle tackled Plato's theory of Forms. In Plato's idealistic view, the ultimate reality is that of Forms, of which tangible Matter, i.e. objects that we can touch and see and smell, are only projections. Aristotle added a variation to that idea, by explaining that the relationship between Forms and Matter was two-way. Notably, since we only experience Matter with our senses, any knowledge of Forms must proceed inductively, by making inferences from observations. The difference with Plato is subtle, but in it laid the germ of experimental science, though it would need almost two millennia and Galileo to blossom into modern Science. The Form for a kind of objects was described by Aristotle with sentences such as "the what it is"; it was translated in Latin into the neologism *essentia*, from which is derived the English term of "essence".

Aristotle was a great fan of classification; he defined *categories* for just about anything he was talking about. While Plato was mostly interested in metaphysics and political virtue, Aristotle's interests extended over natural phenomena, and in particular he dabbled with the classification of living beings. The *History of Animals* (*Historia Animalium* in the Latin translation used by West European medieval thinkers) proceeds from the idea that all living beings that we can observe are physical incarnations of a number of essences, and the categories he establishes really relate to these essences. However, true to his conception of the Matter-Form relationship, Aristotle explains that such

Figure 3.2: Aristotle, the leading influence on natural science in Medieval Europe. This bust is a Roman copy of a Greek original made by Lysippos in 330 BC, so it probably is a faithful representation of Aristotle's likeness.

categories must be worked out from examples, by observation. We do not exactly know how Aristotle proceeded, but from his works it seems clear that he performed many dissections, talked to various craftsmen such as fishermen and bee-keepers, and, in some cases, allowed his rich imagination to fill some gaps in his experimental knowledge. The important part of Aristotle's work is not the exact classification that he came up with, but the method by which he did it. In mythological terms, Aristotle is explaining the way Epimetheus organised his distribution of attributes.

An important idea of Aristotle is that categories should be positive. For instance, he groups together all animals who have red blood, and notices that many of them have lungs as well. Aristotle says that it would be tempting to lump all the others (namely, fishes) into a category of "red blood animals without lungs", but that would be wrong, because not having lungs is not a positive attribute. The right way to classify fishes is that they are "red blood animals with gills". You keep track of Epimetheus's work by the attributes he gave, not by the ones he retained. While the principle is sound, Aristotle himself did not always follow it. For example, after having made a group of animals with red blood, he carelessly stacked all other animals into another group characterized by not having red blood; this is a clear violation of his own principles, and that one had an enduring legacy since two millennia later it engendered the notion of "vertebrates" and "invertebrates". It took centuries for biologists to finally understand that "invertebrate" is no more an adequate method of classification than claiming that objects are either cars or non-cars, the latter comprising teapots, babies and mountains.

Another important idea was that since classification was to be established based on observations, one should first make large families of animals or plants, then find within each family a distinguishing property that allows extracting a sub-group out of that family. This is a top-down approach that promotes positive categories and is likely to yield descriptive groups, thus useful both for thinking and for practical purposes. A bird is a red-blooded animal with feathers and a beak; this is better than saying that a bird is an animal that does not have fur and is not a fish either. In this notion of extraction from big families, or *genera*, based on a specific characteristic, Aristotle can be said to be the precursor to binomial taxonomy, though it would wait for Carl Linnæus to be put to systematic practice.

Another feature of Aristotle's classification of living beings is that he made several classifications, that overlap but do not nest. At one point he lumps together red-blooded animals in one group. Then he also divides animals based on where they live: in the water, on the land, or in the air. Of course there are red-blooded animals in all three areas, and non-red-blooded animals as well. Aristotle does not choose; both categorisations are useful, thus correct. He adds a third based on what the animals eat (herbivores, carnivores and omnivores), which again overlaps the other categories. This means that, as a blueprint for navigating the vast complexity of living things, Aristotle's classification can be quite confusing. It is also very convenient if you want to use Aristotelianism as a complement to another corpus, which is exactly what scholastics were doing. Thomas Aquinas found that the water-land-air classification mapped exactly onto the Creation story described in the Genesis, so that was the basis to be used for subsequent taxonomic efforts.

At the end of the Middle Ages, the Occident intellectual elite had thus adopted the concept of archetypes when it comes to thinking about animals (and also plants), in a "christianised" version compatible with the Genesis, but still derived from the inexpert foundation of our favourite titans. Thanks to the rediscovery of Aristotle and its merging with the strict intellectual discipline of scholasticism, logical tools needed to refine the classification of living beings were available. It only needed some inquisitive minds to pick things up and work from there, which happened at the end of the Renaissance era, in the inception of what would thereafter be known as the Scientific Revolution.

∽ 4 ∾

One Tree To Rule Them All

One fair day in 1582, a young man was in the Cathedral of Pisa, attending mass. That young man was eighteen years old and at that time pursuing studies in medicine. True to his age, he was quite bored with the ritualistic proceedings, and prone to let his eyes and mind wander. His gaze fell upon a chandelier that was swinging from air currents[1]; he noticed that while the amplitude of the swing varied (it increased when a wind came by, then slowly decreased when air was calm), the period appeared to be unchanging. He was way too young and poor to have a portable timepiece on him; the ancestors of pocketwatches were still at that time quite bulky, very expensive, and awfully inaccurate. In any case he would not have been able to consult such an object discreetly; he was supposed to be immersed in the contemplation of the divine miracle of Eucharist, not the motion of furniture. So instead he used his own heartbeat as a crude clock. Later that day, intrigued by the phenomenon, he confirmed it

---

[1]It is not reported where these currents came from, but presumably they happened when someone opened the door.

with two identical pendulums, set into motion at different starting angles. From that moment, the young man was completely smitten with mathematics and physics, then called "natural philosophy". His medical studies were the first casualty of his new interests. He would then, until his death at the age of seventy-seven, be a leading thinker and one of the founders of modern science. His name was Galileo Galilei.

This episode nicely illustrates the differences between the new scientific method, and Aristotelian philosophy. It first begins with an observation without any motive. Aristotle would use observations as part of a mental process that aimed at approaching the essence of things; Galileo was making an observation for its own sake and out of sheer boredom. Then Galileo does something marvellous and truly novel: he makes a measure. He is not content with observing a swinging chandelier; he wants to put it in numbers. This is a direct application of the idea that mathematics are the partition along which the music of the world is played.

Then, back in his living quarters, Galileo goes on with being revolutionary: he sets up an experiment. Implicitly, he assumes that his pendulums could serve as a reasonable mock-up of the chandelier he observed earlier. Making an experiment of that kind makes sense only if you first suppose that you can extract individual attributes from objects and reason upon them separately; in this case, that the swing of the chandelier would come from its length, not from its material, or its position, or its essence as a device to carry candles, and that a length-induced behaviour would appear similarly in a pendulum made from some other material, devoid of any candle, and not hung from the roof of a cathedral. Galileo, in the folly of his youth, is carelessly blasting Aristotle's essences into smithereens.

Last but not least, Galileo's experiment embodies a fundamental characteristic of modern scientific theories, which is their predictive power. Back in his room, Galileo is still clock-less; portable or not portable, clocks are expensive devices and he does not own one, let alone any clock that would count seconds. Instead, Galileo defines a theory, which basically claims that a pendulum's period depends on its length but not on the amplitude of the swing. His experiment will not measure the period; but it will test that prediction, by using two pendulums of the same length but with different amplitudes. The pendulum-length theory posits that both will swing at the same rhythm, so he tries it out. This is how scientific method goes: we make a theory, then we make experiments that exercise that theory, under the hope that we will catch the theory red-handed by the prediction not being fulfilled. There is a famous poster, initially coming from the X-Files TV show in the 1990s, that depicts a photograph of a flying saucer, with the sub-text "I want to believe"; the mantra of the modern scientist is the exact opposite: "I want to disbelieve". A scientist, at the core of his soul, is a merciless killer who hunts down, corners and executes ideas, until one of them survives experiments.

Of course, the 18-year old Galileo was still blissfully unaware of all of this. But he instinctively had the right turn of mind. He dared postulate that experimental data was at least as important as a well-conceived rationale. A famous example involves the way objects fall. Aristotle made the totally obvious inference that since heavier objects are, as it says, heavier, they should fall faster than lighter objects. Galileo tried it; in 1589, according to the biography written by his secretary Vincenzo Viviani, he even dropped two cannonballs of different masses from the top of the famous Leaning Tower of Pisa to test things out. Since Viviani's text was published more than a decade after his death, and more than a century after the purported experiment, the veracity of the anecdote cannot be confirmed. Historians of science usually point out that it would be a lot more convenient for Galileo to make the cannonballs roll over a slightly tilted wooden plank, which would make movements slower, thus easier to measure (remember that he had no accurate clock), and would be way less exhausting than hauling cannonballs up the three hundred steps or so of the tower.

Nevertheless, it is possible that the dropping experiment really took place. Indeed, Galileo, in 1589, had been appointed to the chair of mathematics in Pisa, so he had access to the tower, and it was already leaning (albeit less than today). The story took some momentum later on, one variant involving Galileo dropping the cannonballs just as the University elders passed by, not to knock them out, but to demonstrate his point. This part of the story, while probably fake in its particulars, conveys important truths. The first of them is that Galileo's work was controversial and the subject of disputes that could trigger spectacular antics. Another is that Galileo, while being brilliant in physics, was very disingenuous when it came to human relationships. Whether the cannonballs would hit the ground at the same time or not was not, in fact, the important question. What mattered was the primacy of the experimental data. For a true Aristotelian, the intellectual construction shows that the heavier ball falls faster; if the practical ball does not, then it is the ball's fault. Maybe it was a problem of how it was launched, maybe a question of shape, or material, or any other detail. After all, the observable cannonball is merely a projection of the ball's essence, and that projection can be imperfect, up to the point that the tangible ball would fail to follow the ideal ball dropping speed. For Galileo, this way of thinking must be reversed: if the theory and the practice do not match, then it is the theory that must be amended. In that debate, the falling balls would not prove anything, since the question is not really about the ball speed, but about whether the observation should take precedence over the reasoning.

Galileo's lack of skill at managing the humours of his colleagues was made apparent in the famous controversy about heliocentrism. Galileo had acquired one of the first refraction telescopes, and, since astronomy was part of his attributions, he observed some celestial bodies through it. On the Moon he saw mountains and craters. Around Jupiter he located moons. He even observed sunspots. But in the leading philosophy of his days, inherited from the 2nd century astronomer Claudius Ptolemy[2], who himself had anchored his description of the cosmos in the legacy of Aristotle, the heavenly world could only be "perfect" and ordained as a series of circles, all centred on Earth. The opposition between Ptolemaic geocentrism, and the heliocentric model that posited that planets orbit around the Sun, not the Earth, had been raging for almost a century, since Nicolaus Copernicus had proposed heliocentrism as a mathematical model that better accounted for the observed movements of the planets, with less complexity than the complicated system of nested epicycles that Ptolemy had described. Until Galileo, that specific theme had remained mostly mathematical, and while bitter words had been exchanged, there was no major philosophical issue. In 1609, the German astronomer Johannes Kepler had even described an heliocentric system in which planets followed elliptic trajectories instead of circular. But that was still only calculations. Galileo, with his telescope, was making a much bigger transgression: he was denying the

---

[2]Not to be confused with the eponymous Macedonian general, who got Egypt at the partition of Alexander's empire and founded the Ptolemaic dynasty of pharaohs, that ended at the death of Cleopatra VII and her son Caesarion.

perfection of the celestial world. By seeing rocks and mountains on the Moon, or sunspots, he was turning astronomy into physics. Before him, men just had to raise their eyes to witness divine creations unsullied by the sins of mankind; after Galileo, the planets and the distant stars where no longer the abode of God.

That incorrigible Galileo failed to take the logical escape road, which was to blame the telescope for being flawed and showing illusions. Instead, he began to insist that indeed the bodies in the sky were part of the imperfect, tangible world, and had to be studied as such. Initially, one of the strong supporters of Galileo was the pope Urban VIII; but since the pope's job entailed maintaining a peaceful balance at all levels of the Church, including in its intellectual elites, Urban merely asked Galileo to publish a fair assessment of his heliocentric theory that would also present the then-current official papal view of things, which was Ptolemaic geocentrism. Galileo did so, in the form of a dialogue between three figures; the one supporting geocentrism was called Simplicio, that simply meant "simpleton" to Italian ears. Simplicio was using the pope's own words, and was described as a babbling fool, ridiculed by the two other figures. Quite predictably, Urban VIII was not very pleased at being the butt of the joke, and Galileo ended in a heap of trouble, especially since he had already alienated most of the intellectual world of his time. Amazingly, Galileo appears to have been very candid and never to have given a thought about how bruised egos may react. He strived in academic controversy, and made enemies on a daily basis.

Nevertheless, Galileo still remained a representative of a new way to reason about the world, and a leading force in its promotion. This is what we now call "modern science". The onset of the scientific method was a long process that emerged from a number of important changes over the centuries. We

talked about the scholastics and how they made Aristotelianism fashionable again in the West. Around 1450, Johannes
Gutenberg added some crucial innovations to the concept of
movable type, and thus enhanced the printing press to the
point that it could be used industrially; this greatly sped up the
diffusion of ideas throughout Europe. Improvements in boat
construction, new rigging, and better navigation techniques
allowed explorers to open new horizons, circumventing Africa
to reach India, and also stumbling upon a whole new continent
in the process. Contact with erstwhile completely unheard of
cultures triggered an appetite for knowledge, and a significant
shifting in the perception of the place of Man in the world.
16th century philosophy saw the development of humanism,
which favoured critical thinking and found it fit to consider the
earthly well-being of humans, rather than entirely focusing on
the after-life and possible salvation.

This all culminated in the late 16th and early 17th centuries,
which is the period during which the scientific method was
truly born. Galileo, with his insistence on experiments and
measures, is the proper incarnation of that final step. From
that time, it also began to be possible to think about the
world without doing theology. During ancient Greek times,
a philosopher would make theories about physics that would
encompass everything, up to and including gods. In the Middle
Ages, the Bible was still viewed as the essential source of all
truths, only to be complemented with the inevitable Aristotle
for some technicalities. It took a lot of time to redefine the
scopes of religion and science, notably because the Church,
by definition, deals with eternity, and thus thinks long-term
and is never hasty. This process is now mostly achieved in
the case of the Catholic Church, which talks about morality
and metaphysics, domains which are by definition out of the

scope of science; a strong dualism, that defines the body and
the soul as two separate entities, has helped a lot to reach
compatibility of religion and science. Evolution, cosmology,
and such topics are now considered welcome, if orthogonal, to
religious thinking. The Vatican sponsors scientific-theological
conferences, and the Vatican Observatory is a well recognised
centre for astronomical research.

A fundamental reason for the often perceived opposition
of religion and science is that they exercise human brains in
distinct ways. A religious thinker deals with faith and believing;
a scientist spends most of his time doubting everything. While
both activities are correct in their respective fields, they cannot
happen simultaneously within the same mind. Metaphorically,
religion and science use the same cogs in the brain, but run
them in opposite directions. It thus is very hard to be thinking
about religion and about science efficiently and at the same
time; any switching implies a relaxation time during which
the mind is mostly confused. Some very bright and very rare
people maintained themselves at the bleeding edge of both
theology and physics; the prime example is Isaac Newton. But
for everybody else, specialisation was the way to go: you could
be a full-time scientist, or a full-time theologian, but not both.
This made science and religion compete with each other for the
scarce resource of human attention. It is then no wonder than
throughout history, there was some friction, bitter debates, and
the occasional burning at the stake.

We must note that in the vast panorama of Christianity,
the Catholic Church is only one component, and one that
extremely values intellectualism, verging into utter aridity at
times. There are many other churches, some of them having
spent less time trying to come to terms with science, and still
battling. This is especially visible in the depressing controversies

around creationism and evolution, where both parties appear convinced that religion trumps science or vice-versa. From a strictly scientific point of view, existence or inexistence of God or the act of Creation cannot be proven or disproven; God is simply not defined in a scientific way. From a strictly christian point of view, the same assertion holds: being able, at all times, to doubt the existence of God is a necessary prerequisite of free will; if God cannot be denied, humans cannot be free, only obedient or rebel. In theological terms, one must imagine the Almighty instructing his angels to bury dinosaur bones so that it is possible to think about a world history without needing what Laplace called "the hypothesis of a Supreme Being". Nevertheless, some heated minds seem incapable of accepting that the postulate of universal materialism that is at the root of the scientific method does not equate with the metaphysical assertion that deities do not exist. Thus we will find, pitted against each other, militant atheists invoking science to claim that religion must die, and militant religious people who insist on teaching their own mythology as a scientific theory. This is both bad science and bad religion.

Galileo was not much interested in biology. To witness the new scientific method applied to the natural world, we must jump forward one century, to the year 1707, which saw the birth of two great geniuses. The first one is Carl Linnæus, who in 1761 was ennobled by the Swedish king Adolf Frederick, whereupon he changed his name to Carl von Linné: the "von" is a German mark of ennoblement, while "Linné" is a gallicised version of his previous name. Linnæus was born in Sweden, the son of a Lutheran minister and amateur botanist called Nils Linnæus. Nils himself had chosen his name: as per the old patronymic system of Scandinavia, he should have been "Nils Ingemarsson", since his father was called Ingemar; but upon his admission at the University of Lund, Nils had to show his acceptance of the modern system of family names, and decided to be called Nils Linnæus, out of the Swedish name "*lind*" for the lime tree. Carl inherited that tree-based name.

Linnæus's early studies were meant for a life of priesthood, but it turned out that he was not very good or much interested at such subjects. Instead, he was rerouted to medicine, physiology and botany, the latter soon becoming a life-long passion. In a slightly chaotic curriculum, he ended up in Uppsala University, in which he wrote a thesis in 1729, gave lectures, and published his first books. Crucially, he spent a lot of time looking at all plants that he could find; he also made an expedition to Lapland, the northern part of Sweden, that took six months in 1732. Linnæus was good scientist material: inquisitive, persistent, somewhat obsessive in his attention to details. At the time of Linnæus, the usual nomenclature for cataloguing plants followed the works of Joseph Pitton de Tournefort. Tournefort had made a spirited effort at clumping together information about 7,000 species of plants into a single structured classification, and since he was French, he made sure that it came with a clear and well written narrative. However, while Tournefort's descriptions of individual species were precise and informative, his classification lacked any relation with actual biological realities. Linnæus was dissatisfied with Tournefort, and he decided to invent his own classification.

A number of advances in taxonomy had already been performed since the times of Aristotle. The concepts of species and genus had already been defined, as two separate things; the genus is a group of species that somehow are considered to be in close relationship with each other, that is, closer than with any species that is not part of the genus. As for naming, it had already become somewhat traditional that any species of plant should be

Figure 4.1: Carl Linnæus, ennobled as Carl Von Linné. Oil painting by Alexander Roslin, in 1775; Linnæus died three years later.

named after its genus, followed by one or several words apper-
taining to the specificities of the species within its genus. Since
such names could become long streams of words that really were
descriptions of the species features, there was an ongoing effort
to favour simplified, binomial names: a species's name should
consist of exactly two Latin words, one for the genus, the other
to disambiguate the species within the genus.

Linnæus built his classification on these foundations. He made
several fundamental innovations. The first of them was that all
species should have binomial names, systematically. Moreover,
the species-specific name (the second word) was to be purely
symbolic, not descriptive. This characteristic is a conceptual
leap: it differentiates the naming from the description. This
allows the name to "stick" on the species even when new
information on it becomes available. Then, Linnæus decided
that every grouping, every category, should be defined relatively
to observable characteristics, that all species in the category
share, and none other. For instance, the *Aves* class, for the
birds[3], shall consist of all animals with two feet and two wings,
that lay solid eggs with a calcareous shell, that have a "naked,
extended, toothless jaw" (i.e. a beak), that have warm, dark red
blood, and so on. This is a positive categorisation, as Aristotle
recommended but inconsistently practiced.

Then, Linnæus decided that categories should follow a strict
tree-like structure with exactly seven levels. This means that
categories properly nest, with none of the overlapping that
made Aristotle's classifications so confusing. This is a "tree"
in the sense used nowadays in computer science: a strictly
ordered structure that goes from a single root toward leaves,

---

[3]Linnæus makes science in Latin, like everybody else in his time.

with successive branchings and never any merging[4]. The top category is the *kingdom*; Linnæus defined three, for plants, animals and minerals[5], respectively. Right below the kingdom, one finds the *class*. The animal kingdom is thus declared by Linnæus to contain six classes: mammals, birds, amphibians, fishes, insects and worms. Each class then consists of several *orders*; orders contain *families*; a family groups together several *genera*; and a genus is a group of *species*. The seventh level is for subdivisions within species, and it has several names that depend on the kingdom: *varieties* for plants, *races* for animals[6].

In Linnæus's system, every species has a defined place within the universal tree. For instance, the wild boar (*Sus scrofa*) is in the *Animalia* kingdom, *Mammalia* class, *Bestiae* order, *Suidae* family, and *Sus* genus. There is nothing ambiguous, everything falls into place in an organised and systematic manner. This system is also, by nature, very static. Linnæus was, most of his life, a firm tenant of the "fixity of species". He did recognise that hybridisation could produce apparently new species, but he then explained that the new species were already there *in potentia*, in the preordained list of species that God had established in the Garden of Eden, and confirmed on board Noah's ark. We of course recognise here Epimetheus's figurines, in their christian version. Linnæus viewed his taxonomic work as merely an ex-

---

[4] As per algorithmic tradition, these trees grow "down", with the root on top. It has been suggested that computer scientists have so poor a grasp of outdoors that they don't even know what a real tree looks like. This downward orientation is also the modern graphical layout of genealogies.

[5] Linnæus was really trying to classify the whole of the natural world, not just living things. His classification of rocks and stones has long been abandoned.

[6] Two centuries later, the term "race" acquired a very negative connotation, especially when applied to humans, so biologists now use the much more neutral *subspecies*.

tension of the original "naming" made by Adam, as related in the Genesis.

Linnæus's classification was extremely influential, and is still in use today, with extensive modifications, notably with many extra layers. Yet, despite Linnæus's belief in unchanging species, his classification contained the seeds of ulterior advances that would contradict that very notion of fixity. The use of binomial, symbolic, non-descriptive names allowed thinking about species for themselves, and not simply through their visible features. The taxonomic rank of "family" naturally induced some more or less conscious thoughts about whether some species could have been *engendered*, and maybe begin to construct a sort of narrative for a genealogy of species – this is tantamount of talking about an history of species, that can get spawned along the way, or die.

In 1707, the same year as Linnæus, another great scientist was born. Georges-Louis Leclerc was the son of a minor official, so he was technically a noble. More importantly, he became very wealthy at a very young age through the time-honoured method of simply inheriting from a childless uncle. His father bought in his name an estate, that would later give Georges-Louis his more famous denomination: the *Comte de Buffon*. Once grown up, Buffon would demonstrate both a solid appetite and innate competence for honours, money and science, not necessarily in that order; throughout the course of his life, he would expand his fortune through tactical investments, produce a lot of science, and become quite famous. Buffon dabbled in mathematics and in literacy, both with considerable success; he even joined the ranks of the *Académie française*, the high temple and centre of the worshipping of rhetoric, literature, and the French language, by the French intellectual elites (both at that time and nowadays). But it is in the field of natural history that Buffon left his most enduring legacy.

Through carefully negotiated connections and with the help of his protector Jean-Frédéric Phélypeaux, Comte de Maurepas, then at the (first) apex of his political career[7], Buffon was appointed curator of the *Jardin du Roi*, renamed *Jardin des Plantes* during the Revolution, when being a king was becoming definitely unfashionable. From that office, Buffon promoted work in natural sciences, and himself wrote extensively. His own *Histoire naturelle, générale et particulière* spans an impressive 36 volumes, published between 1749 and 1788 (eight more volumes were published after his death). In many ways, Buffon was an anti-Linnæus. Where Linnæus, immersed in an heavily germanized culture, produced strictly ordained tables and organised lists, the French Buffon wrote everything as a long prose, with many literary flourishes for which he was famous[8]. In the first volumes, Buffon explicitly criticised Linnæus's approach to natural history through taxonomy; the field is called "natural *history*" and Buffon really meant it.

At the start of the 17th century, it suddenly became all the rage, at least in the intellectual circles of Europe, to establish

---

[7]Maurepas was minister of the navy for king Louis XV, and a member of the council of state. His attributions included colonies and seaborne trade, and his unofficial influence was great. He was ousted in a palatial coup in 1749, but made a spectacular comeback as mentor and chief advisor for the young Louis XVI, after the death of Louis XV, in 1774.

[8]It is often reported that the mathematician d'Alembert called Buffon "*le grand phrasier*" ("the great phrase-monger"). The source of this particular piece of gossip proves elusive; it may have been a bit of slander from Jean-François-Joseph-Michel Noël, a professor-writer-politician who made a career in the troubled times of the French Revolution and subsequent bonapartist Empire. Noël was writing from his own recollections, more than a decade after the deaths of both Buffon and d'Alembert.

Figure 4.2: Georges-Louis Leclerc, comte de Buffon. This bronze statue was made by Jean Carlus in 1883; originally located in the town hall, it has later been relocated into Buffon's kingdom, the *Jardin des Plantes*.

a chronology of all events since the Creation. Several leading thinkers tried to estimate how old the Earth was, using of course the Bible as the main source, since it gives detailed genealogies from the first man (Adam) down to king Solomon. Genealogies of kings, starting with mythical and divine figures and ending with current monarchs are a very common fixture of human cultures; you will find such lists everywhere, from the Incas to the Chinese emperors, going through the well-known list of Kings of Sumer. That the Bible includes a similar list of ancestors is not surprising; and when the Bible was written, the concepts of truth and logic as we know them did not apply to the divine world. Reading the Bible in the 17th century and expecting it to be a numerically accurate account was an unreasonable expectation, since its authors did not think that way at all. Nevertheless, so they did, and even big names like Kepler and Newton produced their own estimated date for the initial Creation. They were all within the same approximate range, and, through apparently more chance than rational merits, the adopted conventional date was that proposed by James Ussher, archbishop of Armagh (Ireland): according to Ussher, God began Creation on October 23rd, 4004 BC, at the entrance of the night (thus, around 6 PM or so). 18th century people thus assumed that the Earth was less than six thousand years old.

Buffon said, to Hell with that. From his extensive observations and his quest for an historical narrative, he found that such a young age was simply not compatible with facts. In the *Histoire naturelle*, he described in broad lines an history of the Earth that required its age to be at least ten times as much longer. In a book published in 1778, Buffon estimated an age of 75 thousands of years. This was a compromise: in private correspondence, Buffon admitted that he believed the actual age to be closer to the then-phantasmagoric figure of one million years. For his disregard of Ussher's chronology, as well as for his proposal of a theory of reproduction in which species could be "born", Buffon ran into some trouble with the theologians of the Sorbonne and had to publish a formal retraction. However, he did not remove the offending passages from subsequent prints of his books, and the theologians did not press the issue any further. Buffon had powerful friends, and the king Louis XV, though profoundly christian, was also in relatively poor terms with the clerical authorities because of his scandalously dissolute private life.

Buffon thus considered that species could change over time, albeit within reasonable limits. In his mind, all species could be ranked along a scale of "perfection" and attributed a moral value; of course, humans were at the top. A species could then improve or degenerate over time, or even split into factions with distinct changes. Buffon, for all his genius, was also a child of his time, and thus ascribed to the then prevalent strongly racist views; in his analysis, Caucasian people were "true to their nature" while Asiatic, Africans or Native Americans were all degenerate. Interestingly, Buffon imagined that the climate and the geography had a lot to do with species change. In many ways, Buffon was a precursor of transformism, a theory that later morphed into evolution, as described by Charles Darwin. Darwin himself credited Buffon for being the first to talk about evolution in a scientific manner, but criticised his moral overtone.

In Buffon's works, species are no longer fixed, but they do not have complete mobility either. To keep on with the Epimetheus metaphor, Buffon considered that the clay figurines were in fact a set of small families, with some occasional births, and like in every good family, there were bad elements who strayed into sinful behaviour. Buffon's first foray into transformism territory would prove fruitful, and contagious. Even Linnæus, at the end of his life, appears to have been somewhat contaminated, in that he slightly amended his works to leave open the possibility that, maybe, different species within the same genus could have had a common origin. In a still timid Buffon-like step, Linnæus was toying with the idea that God's creations, and Epimetheus's figurines, might have been the genera, not the individual species.

This was a start. It did not stop there.

Interlude: On Archetypes

In the course of the previous chapters, we tried to follow the filiation of the original idea of archetypes, as applied to the natural world of living beings. The creation myth of Prometheus and Epimetheus is an early representation of that idea, which was then transformed and adapted into a Christian world that purposely imported Aristotelian philosophy over a substrate of fundamental truths revealed by God himself. Then the Scientific Revolution was born from the mixture, and inherited the notion.

It is worth taking a few minutes to review the particulars of that convoluted process. There are two main concepts in archetypes when applied to plants and animals. The first of them is that all individuals are tied to an invisible but powerful ideal essence, of which they are merely projections. Each beast or vegetable is part of its "kind", and only these kinds matter. Individuals may thus exhibit different features, but collectively orbit around a master template. All the thinkers and scientists that we have evoked so far have equated these archetypes with *species*. But what is a species? To put things bluntly, it is a full can of worms. We will talk at length about the definition of species in a later chapter, because we still need a few extra notions for that discussion. But we can already remark that all the efforts of taxonomists, from Aristotle to Linnæus and Buffon, rely on the assumed existence of archetypes: this allows them to use species (or, in some cases, subspecies) as the elementary unit for their intellectual constructions.

The second main concept is that archetypes are fixed: their attributes have been ordained by the gods (by God in the Christian version, by the inept delegates of the gods in the older Greek myth), and they do not change over time. All the individual species have thus existed for all times, and will keep on doing so. Individuals may somehow err from their true blood, but generations after generations must get back to the unwavering reference of their archetype.

These two concepts resonated with the way society was organised in ancient times. In 17th century Europe, any individual was, primarily, a representative of his lineage; the interests of the group trumped that of the individual. This explains the complicated matrimonial strategies by which families were trying to improve their social rank in the very long term. Moreover, the inherent virtues of a truly noble lineage were supposed to stem from the founding ancestor. Subsequent generations could dilute the blood through careless unions, but the central noblesse was a fixed undrifting beacon that merely awaited the right heir to become apparent again. These ways of thinking about social relations and human nature were similar, in their mechanics, to the intellectual construction that spawned Linnæus's classification. It is thus maybe unsurprising that archetypes received their first blows in 18th century Europe, a period later called, somewhat pompously, the "Age of Enlightenment". Philosophers of that age were putting a lot more emphasis on the individual, and disregarded the importance of lineages, viewed as archaic remnants of an obscure past. Though most of such ideas were fueled by dissatisfaction with the political structures of the time, the conceptual shift still had long-ranging direct and indirect consequences for all fields and in particular the burgeoning science that would come to be called biology[1].

Buffon's theories were contradicting the second subconcept of archetypes – i.e. their unchanging nature. But Linnæus's classification might have struck a deeper impact, in that it organised the living species based on their apparent features, thus instilling the idea that what matters is not the essence of the species, but what they look like. From that point, making the hypothesis that maybe the essences could be simply disposed of, and ignored altogether, was just a matter of time.

---

[1] The very term "biology" was coined by none other than Linnæus, though it waited a few decades before becoming the accepted designation for the study of Life.

∽ 5 ∾

The Uncle We Do Not Talk About

In the south of modern Netherlands lies a town called Maastricht. To many European ears, that name is inextricably linked with the Treaty that was signed in 1992 between the members of the European Community, and that created the European Union and its currency, the Euro. The ratification process was difficult and revealed profound divisions within political parties, in particular in Denmark, France and United Kingdom, and contributed to give a bad reputation to that otherwise fine city: for half the electorate base, the name "Maastricht" conjures up visions of decline and loss of sovereignty. To readers versed in palaeontology, Maastricht makes them think about the Maastrichtian, a geological age named after that town because outcrops from that period were first identified in a neighbouring quarry. The Maastrichtian spanned over the last six million years of the Upper Cretaceous and culminated with the K-Pg extinction event, i.e. the wiping out of dinosaurs, probably in the aftermath of a powerful asteroid impact. A dinosaur-killing inferno is not a much better association than a common currency.

In 1764, and then again after 1770, in the Maastricht limestone quarry that ultimately gave its name to a geological age, very strange rocks were found. They looked like pieces of skulls from some great animal. At that time, a few science inclined people were collecting fossils, mostly remains of ancient mollusks, without necessarily considering them as real traces of dead animals. They were curiosities, often viewed as rocks that mimicked lifeforms. Johann Leonard Hoffmann, a local surgeon, was such an amateur collector, so he took interest in the discovery, and sent letters describing the finding to several leading naturalists and geologists, in particular Petrus Camper, then incumbent of the chair of anatomy, surgery and botanics at the University of Groningen (in the north of Netherlands). The skull was very puzzling, for several reasons.

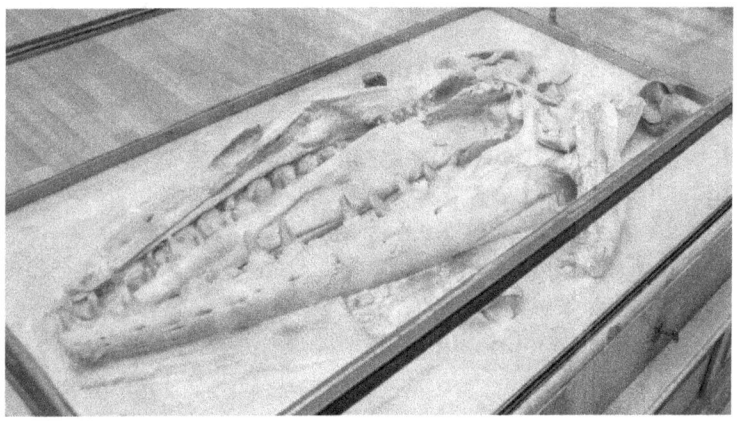

Figure 5.1: The Beast of Maastricht (fossil skull of *Mosasaurus hoffmanni*), on display in the *Jardin des Plantes* in Paris.

The first mystery was the question of how a skull could become embedded in rock. Obviously, the concept of pouring plaster around an object was not unknown, but the block of limestone in which the skull was found was extracted from the earth, and fully natural. Also, the skull was somehow "petrified": the bone had been replaced by another type of mineral. The processes of sedimentation and fossilization were completely unknown at that time, and nobody was even conceptualising the idea of formation of rocks. Like species, rocks were supposed to be eternal and the very image of fixity – as it were, rocks are rock solid.

Even assuming that a dead animal could, through an unspecified feat of involuntary chemistry, find itself in the heart of a stone block, an additional puzzlement was the nature of the animal. Petrus Camper, through comparisons with bones from many other animals, came to the conclusion that the unidentified beast was undoubtedly a marine animal, and probably a kind of sperm whale. But Maastricht was more than a hundred kilometres from the nearest sea. A whale crawling such a distance, to finally die in the middle of a field, was completely unheard of. Camper's identification was to be disputed later on, notably in 1799 by Petrus's own son Adriaan, who recognised the skull as reptilian, close to monitor lizards. But it would have been a really huge lizard, at least five metres in length, and there was a definite lack of gigantic lizards in the Dutch neighbourhood. The closest reptilian beast of that size would have been a crocodile, but that was in Egypt, many thousands of kilometres away, and anyway the skull was much more lizardian than crocodilian.

In 1794, the French Revolution was ongoing, and, after a few years of drafting constitutions and electing assemblies, they ended up with beheading their king, and declaring war to about all the other european countries. And, to the amazement of military headquarters everywhere, they appeared to be winning. As a side-effect of a streak of victories, the town of Maastricht was occupied, and the French troops took hold of the finest of

the two specimens of the beast[1]. The skull was brought back
to Paris, in the *Jardin des Plantes*, where it still is[2]. There, it
fell under the eyes of Jean-Baptiste Pierre Antoine de Monet,
Chevalier de Lamarck.

Lamarck was a talented collector of plants, and a skilled ana-
tomist when it came to mollusks. Buffon mentored him and
scored him a seat in the *Académie des sciences* (which was less
prestigious but more scientific than the *Académie française*). In
1788, just after Buffon's death, Lamarck obtained a salaried position at the *Jardin du Roi*. In 1790, he proposed to change the
name of the garden to *Jardin des Plantes*, a smart political move
that enabled him to keep his head attached to the rest of his body
throughout the troubled months of the *Terreur*, despite his noble status. Through a lot of observations of mollusks, Lamarck
was more and more convinced that species were not fixed. The
beast of Maastricht demonstrated an extreme consequence of
such transmutation of species, as the concept was called at that
time: if species could change over time, they could also die out,
a very novel idea that was later dubbed *extinction*.

Lamarck built a new theory of evolution, and began publishing
it from 1802 onward. The gist of his idea is that species change
over time, through two natural forces. The first is a sort of cos-
mic tendency toward "complexity". Lamarck believed in a natural, physical force that would push everything toward increasing
order. His notion of what constituted "order" is not extremely
clear, but human beings were of course near the top of that complexity ladder, while the worms and mollusks that he so much
studied were close to the bottom. Lamarck was not religious,

---

[1]Rumour has it that the skull was initially hidden, but its location was
revealed after a reward of 600 bottles of wine was offered.

[2]In 2016, the Netherlands government is still officially requesting that the
French give it back.

and did not see that force as having a purpose. It was just, in his view, a physical phenomenon that was to be recorded and possibly measured.

The second force that changes species is adaptation to the environment. Lamarck postulates that in an organism, organs that get used a lot will grow and gain in power, like the biceps of an oarsman, while organs that are not used will shrink and wither, and maybe disappear altogether. Furthermore, by some unspecified mechanism, the offspring of a creature who trained and untrained its organs would somehow inherit the modifications to the organs. The classic example would be the giraffe: Lamarck presumes that giraffes had initially much shorter necks, but they strained to reach the upper leaves of trees, to the point that after a few years they had slightly longer necks; and then, their babies would be born with themselves a longer neck, that they would further elongate through use. After a number of generations, giraffes obtain the very long necks they bear today. This concept is called "inheritance of acquired characters", also known as "soft inheritance".

We now know that Lamarck's ideas were mostly flawed. His notion of complexity could not be defined in any scientific way; for example, when DNA was discovered and understood as the prime mechanism for heredity, it was also noticed that frogs have about ten times as much of the stuff as humans. Therefore, either DNA size and thus chemical complexity was not a good predictor of Lamarck's increased order; or we had to admit that frogs are much higher than humans on that specific scale. Scientists preferred to drop that idea of complexity, since it was not correlated with measurable quantities. For all intents and purposes, Lamarck's complexity was really Lamarck's wish to see human beings, and possibly Lamarck himself, at the apex of the natural world. In that sense, it was Buffon's scale of perfection disguised as the result of a blind and purposeless natural force.

The soft inheritance also failed to be confirmed. In fact, soft inheritance is what happens to *cultures*: behaviours can be acquired by individuals and passed on to their descendants through a process known as "teaching". Humans do that, both consciously and unconsciously. Transmission of some behaviours is also observed in many animals who take care of their children; baby tigers learn the art of hunting from their mother, but do not possess it inherently[3]. But for organs such as giraffes' necks, there is no way any kind of physical training will pass on to the next generation. Your babies won't have a huge right arm because you play a lot of tennis; at best, you could contaminate your children with an inordinate love of the game.

Lamarck still had a profound influence on ulterior naturalists, and in particular Darwin. Lamarck can be credited with having presented the first scientific theory that describes species change as something that depends on the environment, and where inheritance of characters by individuals from their parents is the principle mechanism of perpetuation of a species. This element is the negation of an archetype: animals are part of a species not because they tend toward being similar to an ideal template, but because they look like their parents. A species has been turned into a habit, that sticks and is passed from generations.

---

[3]Mostly. South China tigers, born in captivity in South Africa, are currently being tested for their ability to hunt, to check whether they could be reintroduced in wild areas in China. The test consists in leaving them in a very large area with preys, and giving them food only if they consistently fail to obtain some by themselves. Some tigers, but not all, appear to catch on and learn to be decent hunters by themselves, but it takes some time. Some others are really hopeless.

A recurrent problem for early evolutionists was the undeniable fact that while their theories were all fine on paper, evolution did not appear to be happening in any visible way. Looking around themselves, naturalists would see the same species as their grandfathers, and even Aristotle's descriptions were in phase with animals that were still running, swimming or flying more than two thousands of years later. If species changed in the past, why did they stop? It made no sense that on the span of less than six thousand years allotted to the Earth, all the evolution would have occurred in the first four thousands and then stopped suddenly, just when people were beginning to look at it seriously. And if evolution was still ongoing but so slow that people do not notice it, then six thousand years were way too little a time for evolution to have any significant effect. Evolutionists thus required extra time. They also needed extra proof of species changes. They got both from the opposition between catastrophism and uniformitarianism, championed by Georges Cuvier and Charles Lyell, respectively.

Georges Léopold Chrétien Frédéric Dagobert Cuvier[4] was born in 1769 in the city of Montbéliard (in the east of France). At the age of 10, he first met natural history through books, in particular Buffon's *Histoire Naturelle*, of which he became an assiduous reader. After brilliant studies in Stuttgart, he was short on financial resources and had to find a job; this lead him to become the tutor of the son of a Count in Normandy. By pure chance, he then met a famous agronomist, Alexandre Tessier, who was hiding in the nearby town because of the climate in Paris, which was at that time very unhealthy: you could lose your head if you were not on good terms with the political coterie *du jour*. Through

---

[4] According to other sources, he might also have had "Jean" and "Nicolas" as names; his parents must have been very anxious about him running out of denominations.

the connections of Tessier, when spirits cooled down a little in the capital, Cuvier could begin a fruitful scientific career, notably becoming a professor in 1802 at the *Jardin des Plantes*.

Cuvier was soon hooked on comparative anatomy, a field that consisted mostly in staring at bones all day long. Through systematic and rigorous analysis, Cuvier was able to demonstrate that the African and Indian elephants were two distinct species; and, crucially, that some giant bones that had been found in Ohio[5] corresponded to neither. Cuvier thus surmised that the bones belonged to a member of an extinct species. In 1806 he would name that species the "mastodon", though its binomial name is now *Mammut americanum*[6]. Cuvier also analysed the beast of Maastricht, and confirmed the diagnosis of the younger Camper: the beast was really a kind of lizard[7], and as extinct as mastodons. Cuvier also found about a third extinct species, *Megatherium americanum*, a gigantic ground sloth whose remains were discovered in Paraguay.

With such a detailed knowledge of at least three extinct species, Cuvier formulated a scientific hypothesis that became known as catastrophism. In a nutshell, catastrophism explains that the Earth is normally as serene and well-behaved as can be observed in everyday life, but occasionally experiences severe upheavals, in particular great floods, in which large geologic formations can be carved or altered, and whole species may go extinct. While

---

[5] That is, in the valley of the Ohio; there was no such named state in America at that time.

[6] Names can be confusing at times. The mastodon is from the genus *Mammut*, not to be confused with *Mammuthus*, that pertains to mammoths, who are only distant cousins of mastodons.

[7] Strangely, Cuvier did not name it; the beast acquired its genus name of *Mosasaurus*, as "reptile of the Meuse River", in 1822, and the binomial name was completed in 1829 when Mantell called it *Mosasaurus hoffmanni* to honour the naturalist who first described it.

Cuvier himself avoided any reference to metaphysical concepts, catastrophism was well received by religious thinkers, in particular because the Bible contains a prominent account of a catastrophic flood. Catastrophism was a very neat idea to reconcile the fossil record of animal species that were no longer around, and a fixity of species resulting from the perfection of Creation: that species would disappear was totally acceptable if such an event occurred because of a cataclysm ordained by God himself. Cuvier, though, speculated that such catastrophes would occur "naturally" and on a regular basis, over a long history of the Earth, that had to range in at least the millions of years. Buffon's disregard of Ussher's chronology had not gone in vain.

Catastrophism was the leading theory for a couple of decades, then it met its nemesis, uniformitarianism, championed by Charles Lyell. The son of a Scottish lawyer, Lyell was born in 1797, and was sufficiently well-off throughout his life that he could dedicate his existence to science, and in particular geology. A true precursor, William Smith, had spent much of his life as surveyor and drainer walking the fields of England, and making an extraordinary amount of observations all over the country. Among Smith's remarkable intuitions was the understanding that depth was age: rocks were created through accumulation from the top, so the deeper you dig, the older the rocks you find. Moreover, fossils could help in following such strata: all rocks containing fossils of the same exact species of mollusk are reputed to be part of the same strata. It was a great chance that a keen spirit such as Smith found himself surveying and draining in an area where this mechanism was most apparent: the heart of England is made of sedimentary rocks from the Mesozoic era, accumulated in a shallow marine environment from dead animals with shells, including a large variety of ammonites. Ammonites were vaguely similar to today's nautilus, with spiral shells that had highly recognisable details that differed with each species.

Smith published the first true geological map of Great Britain in 1815. This map would be instrumental in the finding of coal, thereby fueling the industrial revolution. But the scientific consequences were not limited to mining. Lyell took note of the sedimentary process of rock formation that was so obvious from Smith's observations; that process was gradual and slow, and thus completely incompatible with Cuvier's floods. Lyell's reflections led him to propose uniformitarianism in 1830. The core idea of the new theory was that geological laws and processes in the past were identical to what they are now, and operated at the same pace. Thus, changes are necessarily very gradual and take a lot of time. The body of evidence for uniformitarianism was impressive, such as the large strata of sedimentary rocks that Smith had painstakingly tracked from Cornwall to Scotland. A consequence of uniformitarianism was that Earth had to be very old, dozens if not hundreds of millions of years; otherwise, there would not have been enough time to build the mountains and plains and seas that we observe nowadays.

In 1832 Cuvier died, and catastrophism gradually lost its traction. Uniformitarianism was to become the dominant notion of geology, up to the second half of the 20th century, in which it was discovered that though the Earth was indeed very old, and in fact much older than even Lyell assumed[8], it *also* knew its share of sudden events with far ranging consequences, the most well-known being the Chicxulub impact: a really big asteroid collided with the Earth at the very end of the Cretaceous period, and the aftereffects on the atmosphere and the climate may well have been instrumental in the extinction of all non-avian

---

[8]Current estimate is at 4.58 billions of years, give or take 20 millions. We could get precise figures thanks to the discovery of radioactivity.

dinosaurs. Lyell's ideas were quite satisfying to explain rock formation, but did not provide any explanation for extinction of species. Thanks to Cuvier, the phenomenon of extinction could no longer be ignored.

A crucial consequence of Lyell's work is that it became more or less admitted that Ussher had really underestimated the actual age of the world. Thanks to Lyell, deep time was now available, so very slow mechanisms were now fair game. Neither Cuvier or Lyell were very fond of Lamarck's transformism; they believed, basically, in the fixity of species, like everybody else before them – such is the power of Epimetheus. For Cuvier, a species could disappear if you killed them all, which was perfectly within the abilities of a worldwide flood, but apart from such a morbid outcome, species had to remain within their allotted variability range. You cannot escape your true essence. Nevertheless, Lamarck's musing were not lost on everybody; it just needed yet another generation, in this case Charles Darwin.

Charles Robert Darwin was born in 1809. Like so many of early scientists, he could benefit from familial wealth to indulge into studies with no immediate financial return on investment. Initially destined to medicine, he found such studies both boring and distressing when it came to opening patients up with sharp instruments. Instead he spent his days reading and learning about natural history, to the dismay of his father, who tried to fix the situation by sending his errant son to Christ's College in Cambridge. The plan backfired: Charles was sufficiently brilliant to obtain diplomas with little studying, and he made connections that would further his interest in a career in science. In 1831, he obtained a place in a sea exploration that was to last for two years, on board HMS *Beagle* (in fact, it took five years for the *Beagle* to complete its trip around the Globe). Charles's father took some persuading to allow his son to spend years (and money) at the other end of the world, but he finally agreed.

Darwin made many observations during his trip. He confirmed and improved Lyell's theories in the case of the formation of atolls. He experienced an earthquake in Chile and noted signs of the land having visibly raised. He was, like many others before and after him, quite perplexed at the strange australian fauna, in particular the platypus, a kind of rat with a duck beak, that looks so weird that the first dead specimen that reached Europe was thought to be an obvious fake. Darwin's most momentous visit, though, was in the Galápagos Islands, in which he spent inordinate amounts of time observing birds. It turned out that each island had its own species of finch, whose beaks varied in shape. Darwin began to suspect that the beak shape was correlated with the kind of plant food available on each particular island. This triggered a train of thought that was, as trains go, very slow.

Indeed, it took more than twenty years for Darwin to formulate his theory, sharpen his arguments, and write them down in a book. In such an endeavour, he got strong support from Lyell himself (though Lyell was not overly convinced by evolution, he still found the idea fascinating); but what really spurred him into publishing was that a younger naturalist, Russel Alfred Wallace, was having the same kind of ideas and began to publish them himself. As Leibniz had described it, great ideas are like floating in the air; they are the result of accumulated information and theories from previous thinkers, and when you have a novel idea, you can be sure that four or five other people are having the same idea at the same time. The world of science, then and now, was a land of stark competition; you shall publish or perish. Darwin had procrastinated for two decades and it could cost him dearly, so he rushed to write down his masterpiece, *On the Origin of Species*, in 1859. The gist of his new theory of evolution, as it was *not* named at that time, is given in the introduction:

> *As many more individuals of each species are born than can possibly survive; and as, consequently, there is a frequently recurring struggle for existence, it follows that any being, if it vary however slightly in any manner profitable to itself, under the complex and sometimes varying conditions of life, will have a better chance of surviving, and thus be naturally selected. From the strong principle of inheritance, any selected variety will tend to propagate its new and modified form.*

The core of the theory is that it does not assume the existence of archetypes. We can summarise it as follows: children look like their parents. The important part is the one that is missing: children do not look like an unchanging ideal template; they merely tend to have the same characteristics as their direct ascendance. But with generations upon generations, the average features of a species may very well drift. By implicitly rejecting Aristotle's essences, Darwin gives species the ability to vary over time. This is Lamarck's transmutation of species, with an important difference: whereas Lamarck postulated an underlying driving force toward complexity, Darwin finds that there is no need for such a force; one just has not to first assume that there is a force toward fixity. The conceptual jump is immense. Characteristically, all his contemporaries completely failed to notice it.

Darwin's theory, fundamentally, is about following ideas to their conclusions, and thinking statistically. Let's suppose that at some point, through a geological change in the situation of the Galápagos islands, one island becomes split. These islands are obviously volcanic in nature, and a volcanic eruption can perfectly do that. So the two islands start with a mixed population of finches with various beak sizes. However, the two islands may have different exposition to the Sun, or different types of soil, so the plants on both islands soon differ. In the island that bears big cactus, finches with long beaks find it easier to feed; thus, these finches are more likely to lay eggs and have offspring that they will successfully bring to adult age. These young birds, children of birds with long beaks, will also have long beaks. Thus, one generation after the island split, the long-beaked finches have become a majority. The process continues: since having a long beak still confers a feeding advantage, next generation will contain even more long beaks. After a few generations, most if not all the finches on that island have long beaks.

Now, on the other island, things may go differently. For instance, there could be a lack of cactus, but instead a kind of tree with tough seeds. Finches with broad beaks may find it easier to crack the seeds open, and thus have food, with which they gain energy, lay eggs, and have broad-beaked baby finches. As in the first island, after some generations, the average beak shape stabilises, but on a broad beak, instead of a long beak. The two groups of finches, separated by accident and faced with distinct environments, *evolved* into different species.

And that's it. Darwin's evolution theory is just that: from the assumption that there is no mystical force that keeps species similar to a pre-established template, it is only logical that species may drift over time, and when they do, the environment *selects* the features that promote survival. It is important to understand that the whole change is statistical in nature. Lamarck would have imagined that an individual finch concentrates real hard into planting its beak in a cactus, and that would have "trained" it into elongating the beak. Then his children would have inherited the result of such training. With Darwin, nothing of that sort. A finch has the beak that it got when it was born, and there is no amount of pecking that will change its shape. The species change emerges as a census, resulting from the individual probabilities of siring offspring. Some finches may be lucky and others may not be, but well-fed finches have a head start.

Darwin was really thinking things through. That's why he took so much time to publish. With his speciation mechanism by natural selection, and the deep time awarded by Lyell's theories, Darwin could see that many species, maybe all of them, could be derivative from a single root species. In particular, the theory implied that apes really are closely related to humans. Who wants to have a monkey in his family? The uproar was highly predictable. It happened just like Darwin had assumed it would. The controversy raged for decades. In keeping with the tradition of Galileo, Darwin was a bit clumsy in his public relations; in 1871, he published the sequel, that dealt with human beings upfront, and he called it *The Descent of Man*. The term "descent" irresistibly evoked notions of degeneracy and moral decline, which were not at all the intention of Darwin. A 1973 series of BBC documentaries, commissioned by David Attenborough and written and presented by Jacob Bronowski, used the much more motivating title *The Ascent of Man*. Darwin's ideas would have met a smoother acceptance if he had presented them with a bit more of marketing flair.

After much debate, the scientific world recognised that it was a sound theory and that it matched available data much better than the alternative. It was not the end of the story, and a sizable chunk of the general public, being little inclined to science

Figure 5.2: This 1871 caricature of Charles Darwin captures the essence of the controversy on the theory of evolution, which is the idea that apes are part of our family tree – even though Darwin speaks of very gradual changes over many thousands of generations, at which point the notion of "family" hardly makes sense any more.

and in particular mostly impervious to the concept of statistics, remains to this date unconvinced; being kin to a chimpanzee is embarrassing, and they don't want to hear anything of that kind. We will come back to that later. But on scientific grounds, one might conclude that archetypes had received a fatal blow, Darwin having successfully denied them. It was not so. It seems that archetypes relate to a way of conceptualising the natural world that is deeply ingrained in the human psyche. Darwin could not easily let them go, and he postulated, a bit out of the blue, that species change could only be slow and gradual. In this we recognise Darwin's inclination to uniformitarianism, as applied to biology; for geology, he was, consciously, a disciple of Lyell. This arbitrary assertion of slowness allows archetypes to survive: the original clay figurine is now given its own history, but it is still there.

With Darwin's theory, we can still think of living beings as entities that somehow revolve around the template incarnated by the species, and if the change is sufficiently gradual, then it is a reasonably correct approximation to simply say that *the species* changes, and the individuals are merely bound to follow the species. Since the time scales of individual lives and of species changes are (assumed to be) very different, it can make analysis simpler, especially when considering very long period of times, as is typical of palaeontology, to look at species as the interesting unit, and to disregard variations of individuals as mere transient noise. This idea is, in fact, the opposite of what Darwin started with: at the core of his theory, the species are a byproduct of the average features of the individuals; but, as a statistical tool, we prefer to think of the individual as a member of the species. Darwin was feeling this preference, and allowed for it through gradualness of change.

It is maybe not surprising that almost at the same time Darwin was publishing the *Origin*, another influential author, in a very different field, was defining human society as something more than the individuals that constitute it:

> *Society does not consist of individuals, but expresses the sum of interrelations, the relations within which these individuals stand.*

So was writing Karl Marx in 1858. In his social theories, Marx negates the importance of the individual; vast groups of individuals, the famous *classes*, are the interesting unit of action, and humans are merely allowed to be part of their classes and be drawn to their fundamental characteristics. Marx's classes behave like archetypes for humans. This idea was thus well alive and kicking in that mid to late 19th century. Applied to biology, it put back species to the front row. Each finch with a specific beak shape was temporarily permitted to alter the fate of his species simply by its ability to have descendants; once this is done, and natural selection is established as a speciation mechanism, the species takes again the leading role.

In that sense, we may claim that Darwin was the first (with Wallace) to see through the myth of Epimetheus, but long-standing thinking habits got the better of him, and Epimetheus regained his preeminence. The archetypes just obtained a family history.

But what is a species anyway?

✄ 6 ✄

The Ship of the Desert

In 43 AD, the Roman Emperor Claudius had been in charge for two years. Slightly deaf, and with a limp and poor health, he was an unlikely candidate for the supreme office. In his young years, he was often mocked for his intellectual inclinations; he was an amateur historian, and thought a lot about grammar, to the point that he defined three extra letters to be added to the Latin alphabet, to transcribe sounds that he found were poorly served by existing letters. In 41 AD, the current Emperor, Claudius's nephew Caligula, had gone well beyond eccentricity, and was stark raving mad; this prompted a replacement, using the traditional method, which was assassination. The Praetorian Guard, having murdered Caligula, proceeded to ransack the palace and kill a few other persons, as is customary in that kind of chaotic *interregnum*. Claudius, fearing for his life, hid behind a curtain, but did so poorly because he was discovered by a Praetorian soldier called Gratus. For reasons which are still unexplained today, Gratus did not slay Claudius, and instead proclaimed him Emperor. Amazingly, the rest of the Guard rolled with it, and voilà! Claudius was in charge.

Claudius soon took the most urgent decision to maintain peace
and prosperity in the Empire: he made his new letters official[1].
Then, he began to plan for the next logical step, which was
the conquest of Britain. An initial show of force had been
performed almost a century earlier by Julius Caesar, but it
was not followed by actual occupation, let alone colonisation.
Claudius intended to do a proper job, and sent four legions; he
himself joined his troops for the first campaign. According to
Cassius Dio (who was born more than a century after the facts),
Claudius had made sure that:

> *[…] extensive equipment, including elephants, had
> already been got together for the expedition.*

The same Cassius Dio explains that the array of "equipment"
available to troops at that time included not only elephants but
also camels. The classicist and poet Robert Graves thus inferred
that camels were brought along with the elephants. Bringing
desert-adapted camels to the swampy environment of Britain
might look like a weird idea, but for the curious fact that horses
appear to be mortally afraid of the smell of camels. This particu-
larity had been put to use in battle by Cyrus the Great, founder
of the Achaemenid empire; in 547 BC, near the town of Thym-
bra, he assembled a camel corp to fight against the numerous
cavalry of Croesus, king of Lydia, whom he soundly defeated.
The Greek historian Xenophon related that piece of innova-
tive warfare; the scholarly Emperor Claudius had certainly read
Xenophon, who was (and still is) a classic.

Camels come in basically two kinds: with one or two humps.
One-humped camels are called dromedaries; they are nowadays

---

[1]The letters made it to a number of engravings as part of imperial mon-
uments built during Claudius's reign; but they soon fell into disuse after his
death.

entirely domesticated, which means that there is no wild dromedary, and they reproduce under human control as part of an activity known as husbandry, just like cows and sheep. Dromedaries are found from North Africa to Iran and Afghanistan, in a large band that goes from the Mediterranean to as south as Somalia. Dromedaries are the kind of camel you would encounter in Arabia, and if you saw the classic movie *Lawrence of Arabia* then you saw a lot of dromedaries (and one magnificent Anthony Quinn). Linnæus named the dromedary *Camelus dromedarius* in 1758. Dromedaries' foetuses actually have two humps; in the adult, the front hump is only vestigial so one sees, from the outside, a single hump.

The two-humped camel was originally named by Aristotle, who called him the "Bactrian camel". Bactria was an ancient denomination for the area around the city of Bactra, now Balkh, in northern Afghanistan. At the time of Aristotle, Bactria was on the north-eastern boundary of the Achaemenid Persian empire, that Alexander conquered. Bactrian camels are now domesticated, and were already at that time, but it seems that camel husbandry was not that much practiced in Bactria. In fact, Bactrian camels were *imported* to Mesopotamia through Bactria, from areas located further East. The reason for importing Bactrian camels, instead of using dromedaries which where already there, seems to be related to organised, almost industrial hybridisation efforts, which we will describe shortly. Linnæus did not dispute Aristotle's denomination, and named the Bactrian camel *Camelus bactrianus*.

Dromedaries are gracile, while Bactrian camels are robust. The two terms "gracile" and "robust" relate to a dichotomy that was noticed by early scholars in comparative anatomy such as Cuvier. It so happens that when there exists a species of animal, there often exists a second, very similar to the first, except that

Figure 6.1: Two dromedaries. Notice the single hump and slender build. The feet are split into two toes, the bottom of which being covered in a leathery pad, instead of hooves.

Figure 6.2: A Bactrian camel. Apart from the extra hump, that camel differs from the dromedary by a bulkier body shape, thicker and shorter legs, and longer hair, especially on the neck – although the amount of hair varies with the environmental conditions.

one will be "burly", with strong bones and heavy muscles, while the other one will have a much lighter build. For example, the chimpanzee and the bonobo are the robust and gracile species, respectively. This trend has been noticed in many species of vertebrates, even for humans, modern humans being the "gracile" counterpart of the now-extinct robust Neanderthalians. In Cuvier's times, this was just puzzling. In Darwinian terms, this is explained in terms of survival efficiency: both the gracile and the robust form have an advantage over specimens with a more balanced morphology. The gracile form tends to be swifter, thus evading predators or chasing prey more easily; it also needs less food. The robust form is stronger, thus better equipped to fend off predators or to beat off competitors for access to females. Thus, any given group should "naturally" drift to either a gracile or robust build, or possibly to both, leading to a ultimate split into two distinct species.

Dromedaries and camels are large animals. An adult male dromedary may weigh up to 600 kilograms or so. Owing to its robust build, a male Bactrian camel may be stronger and heavier, up to one tonne for the largest specimens. However, if you crossbreed a dromedary and a Bactrian camel, you obtain an hybrid camel that is even larger and stronger than both its parents. The hybrid is also often quite tame; its strength and its docility combine to allow for a great load-bearing capacity, up to twice that of a dromedary. Therefore, hybrid camels have been used in armies since at least the middle Assyrian empire, at the end of the second millennium BC. As recently as the siege of Vienna in 1683, the Ottoman Turkish armies were using hybrid camels for transport and occasionally for charging, there again in hope that it would strike enemy horses with fear[2]. Hybrid camels are then the camelid equivalent of the mule, which is obtained from a male donkey and a female horse; the mule has been the favoured way to haul baggage in the US army prior to mechanization[3].

We do not really know why the hybrid camel is so big and strong; neither do we really understand why the hybrid's children are a lot less useful, in that they are smaller and foul-tempered. Weird size changes have been observed in many cases of hybridisation. For example, if you crossbreed a lion and a tigress, you get a liger, which is the largest living cat; a non-obese liger can exceed 400 kilograms[4]. However, the offspring of a tiger and a lioness is a tigon, which is smaller than both lions and tigers. Ligers and tigons are usually infertile, but not always. In the 1970s, a female tigon mated with a male lion in a zoo in India near Kolkata; of the seven resulting "litigons", some were about liger-sized. Contrary to these big cats, hybrid camels are normally perfectly capable of spawning offspring, but, since this is all under human control, this does not happen because the result is disappointing, economically speaking.

Hybrid camels have a single hump, but often with an indentation that makes it look like it wishes to split into two humps.

---

[2] In 2006, the preserved skeleton of such an hybrid camel has been found during an archaelogical dig near Vienna, amid preparations for the building of a new shopping centre.

[3] Between 1856 and 1866, the US Army ran an experimental program of using actual camels in the arid south-west states; it imported a few dozen camels from Egypt, including an hybrid. The program was abandoned due to lack of political support, since its champion was Jefferson Davis, who sided with the confederates during the Civil War.

[4] Since tigers and lions do not have overlapping ranges in the wild, ligers occur only in zoos, where sedentariness and abundance of food often lead to obesity – just like humans.

In that sense they seem to be an intermediary form, which would be completely viable if only they were given the occasion to grow and reproduce by themselves. The dromedary and Bactrian camels are still regarded as two distinct species, and hybrid camels are "just hybrids", not a species; they are not awarded their own binomial name.

Camels are an interesting example for the concept of species, and what came to be known as the "species problem". In early times, before Buffon, species were easily understood as another name for the archetypes. Each living entity was part of a well-defined category, and though it could take some effort to actually discern to which category the critter belonged, it was still assumed that there was one without any ambiguity. Hybrids were therefore anomalous, but they could be explained away by being created through human agency, and being transient in nature; after a few generations, hybrids would either be swallowed back by one species through further hybridisation, or simply would die out. For the extreme cases, ancient Greeks could resort to the myth of Chimera, a monstrous entity depicted by Homer with a lion's head, a goat's body, and a snake's tail. Chimera was slain by the Greek hero Bellerophon, himself a probable demigod, son of Poseidon. Bellerophon simply shot the beast from above, because he was riding Pegasus, itself an interesting case since Pegasus is a flying horse, thanks to its large feathered wings; Pegasus thus has six members instead of the usual four for land vertebrates, commonly known as "tetrapods" for that exact reason. Since in the divine world composite animals could exist, then meeting a few in the tangible world was deemed possible and unproblematic, philosophically speaking, as long as it did not happen too often. Gods are allowed to meddle with mortals when they see fit.

After Darwin, species become a lot more complicated to define. As we saw, Darwin's uniformitarianism made him define speciation as a slow and gradual process; this allows rehabilitating archetypes as a template for individuals, but a moving, changing template. A fundamental part of a Darwinian species is that members of a given species mate with each other; this is how many variants are born, to be pruned through natural selection. This idea has been formalised by biologist Ernst Mayr in 1942, into the Biological Species Concept. The BSC can be viewed as an operational criterion for determining if two populations are from the same species or not: if they can interbreed, and the offspring are themselves fertile, then they are the same species; otherwise, they are not. The BSC makes a lot of sense from the evolutionary point of view: if two populations can interbreed, then they will do so, and all possible combinations will be explored; natural selection will then apply to the result. But if two populations cannot interbreed, then their evolutionary paths are distinct and there is no particular reason they would evolve in the same ways.

One must mind details: by the BSC, dromedaries and Bactrian camels are *not* the same species, even though they can produce hybrids that are themselves fertile. The reason is that they will do so only if brought together explicitly; their ranges do not overlap. If they don't meet, they cannot mate! Of course, with domesticated species, the notion of "range" must be a bit adapted. But the fact that the Assyrian king Assur-bel-kala, in the 11th century BC, had to send merchants to acquire Bactrian camels for hybridisation purposes, implies that dromedaries and Bactrian camels, at least at that time, were not commonly found in the same areas.

We are not finished with camels. In 1878, the plot thickened, when the Russian explorer Nikolay Mikhaylovich Przhevalsky encountered a few populations of wild Bactrian camels. Since they were a bit smaller than "normal" (domesticated) Bactrian camels, and with a slightly different body shape, Przhevalsky found it fit to declare them a different species, with the name *Camelus ferus*. There are nowadays very few wild Bactrian camels left; they live in a few spots in some very desertic and arid areas in the Xinjiang and Gansu regions in the West of modern China, and in southern Mongolia. A 2009 study estimated the total number of living wild Bactrian camels to be about 900, making them a critically endangered species as per the classification of the International Union for Conservation of Nature. Remaining wild camels are hunted for subsistence and occasionally for sport, but the greatest threat to them appears to be... the domestic Bactrian camel. Shepherding in the Xinjiang being what it is, if a herd of domestic Bactrian camels is brought in the vicinity of a population of wild camels, interbreeding will occur, at which point the wild camels are no longer wild: they have just merged into the domesticated species. One could *technically* wipe out the whole *Camelus ferus* species, i.e. drawing them to extinction, simply by moving around about four or five herds of domestic camels, without killing any of the beasts or even making any of them suffer in the slightest. In fact, the individual wild camels, if they could speak, might well find the arrangement to their collective liking since it would be about providing them with more potential mates.

Whether wild Bactrian camels are a distinct species from *Camelus bactrianus* has long been debated, precisely because biologists felt that status as a separate species should not depend on whether a couple of camel breeders found a truck and decided to move a few hundreds of kilometres. Genetics came to the rescue: now that we can relatively easily decode DNA sequences, it has been possible to compare samples from the skin of dead wild camels with DNA from domestic Bactrian camels. These comparisons found a 3% base difference, which is a relatively high number; a similar difference level is found between, for instance, humans and chimpanzees. It was thus *strongly felt* that wild camels deserved their own species. It shall be noted that using DNA base differences is not at all the same criterion as the BSC; in fact, there are many (several dozens) distinct concepts for species, and that wide plurality is what makes the species problem a problem. Anyway, in 2003, the International Commission on Zoological Nomenclature ruled that, in accordance with the majority (not unanimity) of researchers on that subject, wild camels were indeed *Camelus ferus*, a specific species, and not *Camelus bactrianus ferus*, a wild subspecies of the Bactrian camel.

There is also the matter of the tree. Since Linnæus, we are accustomed to the idea of species being part of a tree-like classification, and, since Darwin, that tree is chronological, like a family tree: each branching off is a speciation event, a sub-population that drifted enough to be considered a new species. Species could die off; this is the extinction phenomenon. However, biologists tend to feel that branches should not *merge back*. If two distinct species are allowed to merge again through hybridisation, then the tree is no longer a tree, in the sense used by computer scientists. This is definitely unclean and inappropriate; in the metaphor of species as a big family of archetypal clay sculptures, such an event would be bordering on incest. If wild camels cease to exist under their current definition, many biologists would prefer to claim that they went extinct, rather than deal with the concept of species mergers.

I wrote earlier that there are no wild dromedaries. This is true from a certain point of view only, because there are "feral" dromedaries. Between 1850 and 1900, at least fifteen thousand camels were imported into Australia, to serve primarily as transport in the semi-desertic interior. Most of these camels were dromedaries, but a few of them were Bactrian camels. Like many endeavours, it seemed like a good idea at the time. When mechanized transport vehicles became commonplace and were deemed more efficient than camels, camel breeding ceased, and several former cameleers dealt with the disposal of the now useless camels by the simple expedient of leaving the gates open and forgetting about the beasts. The escaped camels survived and thrived and multiplied, and in 2008 it was estimated that as many as 600,000 camels were roaming freely in Australia[5]. This population, in fact *these* populations, because there are both dromedaries and Bactrian camels among them, are technically wild, and separated from other camel species by thousands of kilometres, so any interbreeding is very unlikely. This would make them a new species of camels, or two new species, depending on how you account for the Bactrian camels; since feral dromedaries and feral Bactrian camels have overlapping ranges over Australia, maybe they should be considered a single species, there again a dreaded merging event.

Feral camels did not access to the "new species" status. Instead, they were classified as pests, and culling began. "Culling" is a nice word to mean that the camels are hunted and killed; by 2013, their number had been reduced to about 300,000, half their population of five years before. It is, technically, a massive slaughter, in the strict sense of the term since the meat is exported to several countries, mainly Saudi Arabia. The reason

[5] The first estimate was the very symbolic figure of one million, but it was trimmed down after a careful revision of the census methods.

for this rather harsh treatment is that camels in Australia are considered to be an artificial, man-made import and thus a mistake that should be corrected, especially since the camels have a non-negligible impact on the native species of Australia. This notion of "native species" is in itself a highly controversial one that will deserve more discussion in a later chapter. The summary is that feral camels are called feral and not wild because they were philosophically tainted by their domestication. They are now unnatural, hence fair game. The Wild Camel Protection Foundation, dedicated to the conservation efforts for the wild Bactrian camel, does not see anything wrong with the culling of Australian feral camels, though they requested that the operation be done "humanely". One might argue that there is nothing more human than killing large mammals to eat their meat, since humans have been doing so for hundreds of millennia; the subtlety is probably in the difference between the terms "human" and "humane".

My intention here is not to elicit an emotional response and point fingers at the flagrant injustice against the feral camels. Whether trying to keep a separate wild camel population alive, and whether to reduce or even eradicate feral camels in Australia, are good ideas, depends on a lot of parameters and in particular on a moral framework that can define things as "right" or "wrong". Science is amoral; the notions of good and evil are outside of the scope of science. What science can help you with, is working out consequences: as we saw, the heart of a scientific theory is its predictive power, so science can potentially tell you what will happen if you perform or omit to perform some specific action. It is still up to a non-scientific set of moral values to declare whether these consequences are desirable or not. The example of the feral camels is still an important one, in that it shows that there is a definition issue with the notion of species: the problem is not solved, and, possibly, we humans might collectively hope that it will remained unsolved. International treaties and local laws on conservation tend to be expressed in terms of "species", without a precise definition of what a species can be. This is quite apparent in the case of the wild Bactrian camels: the ICZN 2003 ruling automatically triggered protection measures in China and Mongolia, including captive breeding programs. Such actions would not have taken place had the ICZN decided that the wild camels were merely a subspecies.

This yields yet another definition of species: a species is that which we deem worth conserving. A current trend is in the replacement of the term "species" with the expression "Evolutionary Significant Unit"; the *significance* relates to the worthiness for conservation. The whole topic of conservation raises many questions on what we are trying to conserve and why, and also whether we do it properly. For example, does it really make sense to run a captive breeding program for wild camels? Captive breeding is the essential step of domestication; can camels bred in captivity still be considered "wild"?

The camel saga is not finished. Palaeontologists have worked hard to try to understand the family history of the camelids. From the fossil records and genetic analyses, it seems that the ancestors of modern camelids originated in North America in the late Eocene, about 45 million years ago[6]. *Protylopus petersoni* was about the size of a big dog, and would have looked a bit like a shrunk llama. The group prospered and diversified, with species assuming a large range of sizes, weights and shapes. About 3 to 5 million years ago, continental drifts, vulcanism

---

[6]Thus, 20 million years *after* the end of the big dinosaurs, but still a good 20 million years before the appearance of animals that could be called "apes".

and variations in sea levels[7] opened two "land bridges", one between North America and Asia (joining Alaska with northeastern Siberia, where there now is the Bering strait), the other between North and South America; the latter still exists today and is called "central America" for lack of a more imaginative name. Prehistoric camels crossed these bridges and established new populations in Asia and in South America. The contact with the North American populations was only episodic, so evolution happened and these distant groups drifted apart. The North American camels ultimately disappeared in the so-called "megafauna extinction event" about 13 thousand years ago. The Asiatic group initially consisted of beasts similar to modern wild Bactrian camels; since Asia is large, population spread led to independent groups that kept on drifting, resulting in modern Bactrian camels and dromedaries.

Domestication further altered the species. Of course, a given individual won't change in the course of his life, regardless of how many humans are in its vicinity. However, humans always practice selective breeding, if only when deciding which camel will be used to produce the next generation and which camel will be turned into tonight's meal. So average features are modified upon successive generations, sometimes dramatically, and quite fast by palaeontological standards; next time you meet a chihuahua dog, think that its ancestors were fearsome wolves only a few thousand years ago. Domesticated animals are the most obvious counterexample to Darwin's uniformitarianism. In the case of camels, domestication increased the average beast size and improved their temper, because huge docile beasts are more useful to humans. The same phenomenon applied to cat-

---

[7]A cooling climate triggered accumulation of ice in polar ice caps and even "glacial periods", each of them implying a lowering of the sea level by as much as a hundred metres.

tle, pigs, sheep, and about all other domesticated animals. Even cats, which are usually considered to be close to wild species of felids, show signs of selective breeding, the favoured trait being in that case "cuteness".

The South American group fared differently. It seems that the mountainous environment of the Andes favoured smaller, more nimble individuals, leading to the emergence of at least two genera, *Lama* and *Vicugna*. Nowadays, each genus contains two species, one being generally considered to be the domesticated form of the other. The *Lama* genus contains the wild guanaco (*Lama guanicoe*) and the domestic llama (*Lama glama*). The *Vicugna* genus splits into the wild vicuña (*Vicugna vicugna*) and the domestic alpaca (*Vicugna pacos*). All four species have been used by human populations for centuries if not millennia. William Prescott, in his classic account *The Conquest of Peru*, relates the adventurous, ruthless, murderous and momentous conquest of modern Ecuador, Peru and Chile by *conquistadores* led by Francisco Pizarro in the 16th century; in that book, he describes the South American camelids, calling them collectively "the peruvian sheep" because of their role in the pre-conquest society. The Incas were raising llamas and alpacas for their meat and their wool; they also occasionally captured and sheared vicuñas, whose wool was the rarest, smoothest and most expensive, thus highly prized and officially reserved to the highest aristocrats. Tragically, the Incas did not know the wheel, and thus never thought about using llamas to propel chariots[8].

---

[8]The symmetric situation occurred with Aztecs in what is now Mexico: they knew the wheel, but had no available burden beast larger than a big fowl, so they used wheels only for toys. Both Aztec and Inca empires were fully pedestrian. Their ability to respond to Spaniard invasion in the 16th century might have been a lot different if they had known of more efficient transportation.

Early colonial reports describe an extra variety, called the Chilihueque, now extinct. It is not known whether they were a subspecies of llama, or locally domesticated guanacos, or maybe something else altogether. It is entirely possible that extra species resulting from the diversification of the original southern group of camelids may have survived to relatively recent times. It shall also be noted that llamas, guanacos, alpacas and vicuñas can all breed with each other and have fertile offspring, so by the BSC they should all be part of the same species. They are still considered distinct species, partly because of isolation of different populations, and partly for historical reasons. Even now, a strong force in taxonomy remains: "it was named that way before". You do not go on contradicting Linnæus without robust arguments.

Llamas and alpacas produce substantially more wool than their wild counterparts; they were obviously selected for that trait. They still are rather small animals, so they do not produce a lot of wool. They also are vicious beasts that keep a definite hostility toward humans that generations of selection have not succeeded in removing; they bite, spit and kick whenever they have the occasion. Comparatively, camels from the Old World are tamer, and since they are bigger, they have more surface to grow hair. Unfortunately, these big camels have short and coarse hair. Economically speaking, it would be very nice to have a sort of camel, huge and docile, but covered with the long flowing hair of a llama (or, even better, vicuña hair). From the idea to the realisation, all that was required was a bit of funding. It happened in Dubai, a country that both has dromedaries on hand, and also a lot of cash to spend in order to invent a new future for their society, when their great reserves of crude oil finally run dry. The Camel Reproduction Centre started work on hybridisation between camels and llamas. In 1998, they succeeded: the first "cama" was born from an artificially inseminated female llama[9], the father being a dromedary. Ten years later, four other live camas had been produced.

The new individuals might be fertile; at least they show signs of a willingness to reproduce, but we do not know (yet) if it would work. Their initial production, though, was not easy; other combinations, such as between a female dromedary and a male llama, did not yield viable individuals. Even with a female llama and male dromedary, the success rate is abysmal: it took about forty attempts to obtain the first cama. Moreover, as per the original goals, the cama looks like a spectacular failure: the first cama is not much larger than a llama, and shows all the vindictiveness of his mother's side; apart from morphological peculiarities (in particular in his feet), what he got most from his father was his short hair. That first cama is thus irremediably less valuable than both llamas and camels. Experiments keep on, though, in case other combinations, or ulterior selection, allow for improvements. Also, it may be suspected that while greater economical value was the stated objective, it might have been a simple pretext to get budget, and the real motivation was the fun of making the first cross-continent mammal hybrid. The most stupendous advances in science have often been achieved by scientists who were after pure knowledge, and bullied or swindled investors into providing money for hypothetical returns on investment that failed to pop into existence in an economically reasonable delay.

Camels and other camelids show that there is considerable complexity in the notion of species, and also considerable flexibility in our collective willingness to enforce any strict definition. The BSC, the biological criterion for species, has a number of flaws; we saw some of them in the case of camels. The first flaw seems

---

[9]Some reports claim that the mother was a guanaco, not a llama.

obvious but must be stated nonetheless: it works only for organisms that reproduce sexually. A lot of mono-cellular organisms reproduce by division: an individual is a single cell, and it splits into two brand new organisms. There is no breeding in that case. A variant is called parthenogenesis, in which a single individual makes children all by itself; the New Mexico whiptail (*Cnemidophorus neomexicanus*) is a species of lizard where all individuals are female, and each can lay fertile eggs all by itself[10]. There again, no breeding on which the BSC may be applied.

Another problem with the BSC is that it relies on a predefined notion of population. Indeed, try as you may, you won't get camel calves out of two male camels. Thus, two male camels cannot "breed together" though they may breed with the same female. Similarly, it would be felt a bit unsound, scientifically speaking, to state that camels who are too young or too old to breed are no longer part of the species. The BSC must be understood for established populations: two *groups* are part of the same species if they may interbreed, in the sense that some members of one group can mate with some members of the other group and have fertile offspring. This still relies on the two groups being assumed to be homogeneous, and the BSC will be no help for that.

As we saw with camels, the breeding criterion is at least partly geographical. Dromedaries, domestic Bactrian camels and wild Bactrian camels are separate species per the BSC only because they live in different areas and do not meet. Feral camels in Australia are similarly separated but we would prefer it if they are

---

[10] According to a 2011 study, individuals that are able to reproduce by parthenogenesis can be obtained through hybridisation between two other species. Mammal parthenogenesis was successfully induced in a few species in the lab, but often results in abnormal development and was never observed "in the wild".

not a new species, even if the BSC would imply it. Geographical isolation is a transient condition that can change in the blink of a geological eye, especially when humans are involved; it is, by nature, more catastrophist than uniformitarianist.

Another big problem with the BSC is its inability to handle *time*. Today's camel will certainly not mate with a camel from one century ago, since the latter died way before the former was born. The BSC cannot thus be applied to delimit species chronologically. We cannot assume that any individual always belongs to the same species as its parents, since that would negate evolutionary drift and would bring us back to Epimetheus's sculptures. We really want our notion of species to be able to cover gradual drift: we accept that the boundary between two species is necessarily fuzzy, but we will want to be able to say that after three millions of years, this is no longer the "same" species. This particularly matters to palaeontologists, who always deal with millions of years and can never try to mate two of their specimens together, since any two fossils are very unlikely to be contemporaries, and are reduced to petrified bones anyway, which tends to impair their abilities at engaging in sexual reproduction.

A final issue with the BSC is that it occasionally fails to be transitive. If population A and population B can interbreed, and population B and population C can interbreed as well, then the BSC tells us that A, B and C are part of the same species and population A and C should be able to interbreed as well. We unfortunately encountered cases where this transitivity property does not hold (often with species of frogs, which tend to both have small ranges, and high genetic plasticity). Even with camels, we saw one of the roots of the problem: between "breed" and "don't breed", there are a lot of intermediary situations, including "they can breed but usually don't feel like it", or "they breed but the offspring is less often viable than in intragroup breeding".

In fact, the notion of "species" is increasingly being attacked for being not only hard to pinpoint, but also relatively useless for science. The advent of genetics, in particular the discovery of DNA, has considerably changed our outlook on evolution in the last decades.

৩ 7 ৩

The Blueprint of Life

Between 1856 and 1863, an Augustinian friar called Gregor Mendel (born Johann Mendel in 1822) conducted a series of experiments in the garden of his monastery in the Czech town of Brno. Mendel wanted to explore and measure the inheritability of traits; he initially started with mice, but his bishop found that the close study of animal sex was inappropriate for a friar, so Mendel switched to plants, specifically to peas (*Pisum sativum*), whose seeds were easily obtained and grown in a garden under Czech climate; the produce could also be eaten, so even if the experiments were not fruitful, the endeavour would not be vain.

Mendel was thorough and cultivated about 28,000 plants during these seven years. More importantly, he was inclined to mathematics and statistics, so he categorised, accounted and tallied his plants based on observable traits such as seed shape and flower colour. Using his detailed results, he could work out the basic laws of what would later be called *Mendelian inheritance*. In Mendel's analysis, each trait is triggered by a *gene*, which is a unit of inheritance. Each organism has two copies, called *alleles*, of any given gene. Depending on the gene we are talking about, several kinds of alleles may be in circulation, but an individual has exactly two of them. That organism receives one from its mother, and one from its father. Which one of her pair the mother gives to her child is essentially random, and the same holds for the father; it can change each time (which is why siblings are not identical). Moreover, the choice is more or less independently made for each gene; if you consider two genes, each with two alleles, the four possible combinations of transmitted alleles are equiprobable.

Alleles can be *dominant* or *recessive*. Recessive alleles keep quiet in the presence of a dominant allele. A classic example is blood type. Humans can have a blood of type O, A, B or AB; the type matters when doing transfusions, because the receiver's blood plasma must not consider the imported blood cells as foreign and try to destroy them. The blood cell type comes from a single gene, that has three types of allele: O, A and B. The O allele is recessive, while the A and B are dominant. Thus, if an individual has two O alleles, then his blood type will be O; but if he has one A and one O, then the A takes precedence and the resulting blood type is A. There are thus two ways to have blood type A: with two A alleles, or with one A allele and one O allele. The O allele, in the latter case, is quiescent and does not impact the owner's biology; but it can resurface in the children. For example, a father and a mother both with blood type A may still have children with blood type O: if both are *heterozygotous* (meaning that they have two distinct alleles for the gene), then they both have an A and an O allele, and there is a 25% chance that they both transmit their O allele to their common child. The child will then be *homozygotous* (two identical alleles) and have blood type O. When two dominant alleles are present (e.g. A and B, in the case of the gene for blood cell type), they are said to be *co-dominant* and both will have a biological effect (in this case, blood type AB).

Mendel's results went quite unnoticed at his time. People failed to realise that they were not about hybridisation, but also mattered for all "normal" reproduction. His work was rediscovered in the early 20th century, and after much observation and experiments, the chemical mechanisms underlying genes and alleles have been refined. The genes were pinpointed as being parts of structures located in the nucleus of each cell, called *chromosomes*. Chromosomes are made of a very complicated molecule called deoxyribonucleic acid, in short DNA. The "double-helix" structure of DNA was worked out in 1953 by James D. Watson and Francis Crick, based on X-ray diffraction pictures from earlier researchers. DNA gave us explanations on Mendel's laws, and new tools for thinking about species and evolution, and for investigating the relationship between genera.

DNA consists in two long strands that twist around each other and are linked together at regular intervals. You can imagine the double-helix structure as an hybrid between a ladder and a corkscrew. The twisted ladder poles are made of a type of sugar (deoxyribose) with an added phosphate group on the outer side. Each step in the ladder consists of two *nucleobases*; it is called a "base pair". There are four kinds of nucleobase: adenine (A), thymine (T), cytosine (C) and guanine (G). Only two combinations are possible: adenine with thymine, and cytosine with guanine. If you follow a strand, you will then see a long sequence of bases of the four kinds, and the opposite strand contains the mirror sequence. When geneticists talk about "sequencing", they really mean determining the sequence of nucleobases of one strand, and this is represented as letters such as "TGAACAGCTAG…". In that example, the other strand would then contain "ACTTGTCGATC…".

DNA encodes *proteins*. In the complex chemistry of a living cell, at some times, dedicated enzymes will inspect a chunk of DNA and "interpret" the sequence of bases as an encoding for amino acids[1]. Amino acids are compounds that are assembled by plants and that animals obtain by eating plants or other animals. Proteins are sequences of amino acids; exactly twenty distinct amino acid types are used[2]. The decoding enzyme will map each sequence of three successive nucleobases (a "codon") onto one of the twenty amino acid types (e.g. "AGG" is mapped to arginine) and strings amino acid molecules together, until a "stop codon" is encountered. With four bases, there are $4 \times 4 \times 4 = 64$ possible codons; the *genetic code* may map several distinct codons to the same amino acid. The genetic code appears to be almost universal, with only very slight variants in mitochondria and in some types of monocellular organisms.

Proteins may contain many amino acids; the longest known is called titin[3], a variant of which containing almost 36 thousands of amino acids. Conversely, proteins can be quite short; when they contain less than about 20 amino acids, we tend not to call them proteins any more, but that's more a question of terminology than biology. Insulin, a well-known protein produced in the pancreas in most vertebrates, whose lack is the source of a kind of diabetes, contains 110 amino acids. When a protein is finished assembling, it folds: due to variations in electromagnetic potential at the surface of individual amino acids, themselves due to their individual chemical nature and interactions with

---

[1] I am simplifying things a bit here; in fact, a copy of the strand chunk is assembled as RNA, a molecule that is similar to DNA, and the decoding of bases into amino acids is done on the RNA.

[2] Not *really* exactly. After protein synthesis, other proteins can have a chemical effect on the protein, and "patch it", leading to the inclusion of two extra types of amino acids.

[3] The name is derived from "titan" because of its gigantic size.

their neighbours in the protein, the chain of amino acids rolls on itself into a kind of knot. The chemical activity of the protein will depend on the resulting shape, and on the amino acids that end up on the outside. About everything that happens in a cell is the result of the chemical action of a protein, including cell growth and division, and even the protein synthesis mechanisms themselves: the "enzymes" that oversee the transcription from DNA are themselves proteins. Thus, the base pairs in the DNA really are the partition for the whole music of life in a cell.

It is worthwhile to make a pause here and let the concepts sink in. The body of a plant or an animal consists of cells in various types and numbers; the smallest, microscopic living beings consist in a single cell, but larger animals contain a lot more. A typical human body will contain many; the count has been estimated in 2013 to be about 37.2 trillions (American trillions, i.e. 37.2 millions of millions), though it varies depending primarily on the subject weight. All of these cells contain a nucleus[4], and in each nucleus there is an integral copy of the whole DNA of the individual. All these cells came into being by division, a process by which a cell splits in two, and the DNA from the source cell is duplicated. All this building and divisions is done by the action of proteins that are assembled using the DNA itself as blueprint. There is thus a lot of redundancy in the information content. This also explains why making a clone from a single cell of a source individual is not scientifically unsound: each cell contains all the needed information to build a complete individual.

Mendel's genes and alleles describe DNA behaviour. In a cell, the DNA is stored as big chunks: these are the chromosomes. In a given species, the number and size of each chromosome is fixed. They normally come in pairs of similar size and function;

---

[4]Almost. Some cells, like red blood cells, do not have a nucleus.

the two chromosomes from a pair are called *homologous*[5]. For example, a human cell will contain 46 chromosomes, in 23 pairs. A *gene* is a specific region in a chromosome; for instance, the *INS* gene, that contains the encoding of the insulin protein, is located on chromosome 11, starting at base pair 2,159,779 and of length 333 base pairs (330 base pairs that encode the 110 amino acids, and an extra "stop" codon of 3 base pairs).

Since a human cell has two homologous chromosomes in each pair, then it has two "type 11" chromosomes, hence two instances of the *INS* gene: these are the two alleles for the gene. These two alleles may vary in their exact sequence, thus encoding for different variants of the corresponding protein. Genetic diversity is in these variants: while everybody has two alleles for the *INS* gene, not everybody has the exact same two alleles, and thus their bodies produce distinct kinds of insulin, with subtly different chemical actions within the body. A bit of potential confusion must be dispelled at that point: DNA has two *strands* and these two strands have nothing to do with the two *alleles*. The two strands are chemically mirrors of each other (to each guanine a cytosine, to each adenine a thymine). The two alleles correspond to an area in two complete chromosomes, each with its two strands.

DNA is very long, but cells are small, so DNA spends most of its time completely twisted and folded. *How* it folds depends on its nucleobase sequences, in a way which is not unlike protein folding. Which parts of the DNA are eligible for transcription

---

[5]There are organisms with sets of more than two homologous chromosomes, e.g. each body cell of the Uganda clawed frog (*Xenopus ruwenzoriensis*) has no less than twelve instances of each chromosome. In humans, extra chromosomes may occur as accidents in the first steps of embryo development, often with deleterious effects such as Down syndrome, also known as trisomy 21, which entails an extra copy of the 21st chromosome.

into proteins depends on how the DNA is folded, so a gene or a DNA chunk can have both an action as the template for a specific protein, and as a promoter or inhibitor for the decoding of another gene. Most of the DNA does not take part in protein encoding stricto sensu, but some of that noncoding DNA is still important in that promotion and inhibition role. Further complicating the DNA topology is that decoding can work with either of the two strands of DNA, and in both directions, because DNA is not strictly oriented. The same base pair can thus participate to several genes.

In humans, there are 23 pairs of chromosomes. In total, the 23 chromosomes contain a bit less than 3 billion base pairs, that is, each DNA strand contains 3 billion bases, and the mirror strand has the 3 billion corresponding bases. Moreover, each chromosome is doubled (the two homologous chromosomes) so each cell will contain about 6 billion base pairs. Of these, about 1.5% only is coding DNA, and 7% or so are noncoding but "important" in that modifying them alters transcription of some genes and thus can have an impact on the cell's biological functions. The rest, more than 90% of the human DNA, is dubbed "junk DNA"; junk DNA is very convenient for evolutionary geneticists, as we will see shortly. The 1.5% coding DNA represents between 20,000 and 25,000 genes. If you do the maths, you will find that a gene will consist of, on average, about 2,000 base pairs, resulting in proteins with about 700 amino acids (since it takes three base pairs for a codon).

When two animals mate, the child's DNA is assembled from the chromosomes of both parents. In each pair of homologous chromosomes in the mother's cells, one chromosome is transmitted to the child, and the other is not. The same holds for the father. Thus, the two "type 11" chromosomes of the child will consist of one chromosome from the "type 11" pair of the mother, and one chromosome from the "type II" pair of the father. The meeting place of these chromosomes is the *egg*, a single cell which is the very first cell of the child. All the cells of the child body ultimately come from divisions of that single source cell. The child's genetic baggage is therefore a mixing of half the mother's alleles, and half the father's alleles.

The source cells in the mother and father for reproduction (called the ovule and spermatozoid, respectively, at least in mammals) are special cells collectively called *gametes* that contain only one chromosome of each type, not two. The choice of which chromosome they contain is essentially random; it is done in the last cell division that results in two gametes (this process is called *meiosis*). Since a human cell contains 23 pairs of chromosomes, the number of distinct gametes that it can produce is 2 raised to the power 23, i.e. 8,388,608. The number of combinations from the mother's and father's gametes is thus 8,388,608 × 8,388,608 = 70,368,744,177,664. This explains why getting identical children is rather improbable[6]. In fact, the number of possible combinations is substantially higher than that, because during cell division, the chromosomes may break and be glued back, swapping heads or tails in the process: this is known as a *recombination event*.

Except for recombination events, we can see that two alleles that are on the same chromosome will be transmitted together or not at all to the child[7]. Mendel's second law, that states that alleles are transmitted independently of each other, is only partially true;

---

[6] Except for "true twins" who start as a single egg that, at an early stage of development, splits into two embryos, for reasons which are still unclear. The two twins have exactly the same DNA.

[7] Two alleles on the same chromosome will be transmitted separately only if a recombination occurs between them; thus, the closer the two genes are, the more "linked" they are.

Mendel was mostly lucky that in his peas, the traits he was following happened to be based on genes located on distinct chromosomes.

Every time a cell divides, the two strands of DNA are taken apart from each other, and each strand is complemented with a mirror strand. Thus, each chromosome becomes two chromosomes. The two descendants of the cell get one each. The process is complex since it processes billions of base pairs, thus errors may occur. Among the vast paraphernalia of enzymes in cells, some detect and correct errors; but some errors will get through. This is called a *mutation*. If the mutation occurs in junk DNA, then nothing else happens; that's the point of junk DNA being junk[8]. However, if the mutation happens in a part of DNA that has an influence on the protein synthesis process, then it *may* have a biological impact on the cell functioning. If the mutated cell, or one of its descendants through cell divisions, becomes a gamete, then the mutation may be transmitted to offspring.

Random combinations of chromosomes (including recombinations), and mutations, are the two main mechanisms by which species "try" variants, to be pruned by the unrelenting Darwinian natural selection, or by pure chance in the case of traits that do not confer any particular survival advantage or disadvantage. Mutations, in particular, are believed to occur at a relatively steady rate, mostly independent of the environment. It is theorised that high energy photons, radioactivity, or some chemicals, may increase mutation rate; but mutations do not occur in significant quantities for transient exposures to such mutagen agents, as far as evolution is concerned. A large scale

---

[8]Strictly speaking, junk DNA is junk because its location in the folded DNA happens not to code for a gene and not to influence gene transcription into proteins. Mutations might alter the folding and thus "un-junk" DNA. This is very rare in practice; most mutations in junk DNA have no effect.

experiment in human mutations took place in 1945 in the Japanese towns of Hiroshima and Nagasaki. Atomic bombs produce a lot of high energy photons, free neutrons, and radioactive fission products, that thus can induce mutations in cells. It did increase the rate of mutations, and in particular it triggered many cancers: a cancer is basically a cell that mutated in such a way that it begins to divide anarchically and boundlessly, ceases to perform its normal job, and is not recognised as "foreign" by the immune system. A cancer ends up killing the host by all the cancerous cells that take the place of other cells without ensuring their function in the body.

But it takes a single mutated cell to start a cancer, and the overwhelming majority of the body cells are not mutated. In particular, gametes, which are not numerous in the human body and do not divide often, tend to escape unscathed. The probability of one of your gametes to be mutated remains insignificant, unless a *lot* of your body cells are mutated, at which points you have bigger worries. The bottom-line is that when you get A-bombed, if you are not vaporised by the heat, crunched by the shock wave, or killed by the onslaught of high-energy neutrons, then you may get a cancer, but children you conceive after the blast won't have an extra arm, and they will be healthy[9]. The same conclusions held in the aftermath of the Chernobyl nuclear accident.

All of this chemistry is very useful for mapping evolutionary ancestry. It is often reported that humans and chimpanzees differ by only 1.2% of their DNA. In fact, the exact figure depends on how you count things such as deletions – the total number of

---

[9]Children conceived *before* the bombing may not fare as well; being exposed to fast neutrons is very unhealthy, especially for cells involved in embryo development, because any mutation at that point tends to have consequences for the whole body.

base pairs is not the same in humans and chimpanzees. If you do the accounting in a different way, the difference can be up to 5% or so. The point is that when you choose an accounting method, you can apply it on different species to try to work out the similarities at the genetic level. The numerical value is not important in an absolute sense, but in comparisons: the statistical test that says "1.2%" when comparing a human with a chimpanzee will also return less than 0.1% when comparing any two humans together. Thus, while chimpanzees are declared to be rather close to humans in the evolutionary tree, they are still twelve times farther from you than the most genetically foreign of humans. Similarly, the 3% difference in base pairs between domestic and wild Bactrian camels, when compared with the figures obtained between other species with the same methodology, was found to be significant, i.e. similar to the difference you get when comparing individuals from distinct species. This genetic study was instrumental in the ICZN ruling that the wild camel was really a different species.

This new, genetic outlook on taxonomy turned out to be very powerful when applied to a novel classification method called *cladistics*. Cladistics were conceived as early as 1901 by Peter Chalmers Mitchell, but the term "clade" was introduced only in 1958. A *clade* is the set of organisms that all share a common ancestor, so it includes not only that ancestor but all its descendants. Cladistics is the method that produces *cladograms*, which are trees of clades (clades that contain sub-clades, and so on). *Phylogenetic nomenclature* gives names to clades. For any two organisms, it is conceptually possible to walk back to their last common ancestor, and thus define as their common clade the set of organisms that also descend from that common ancestor.

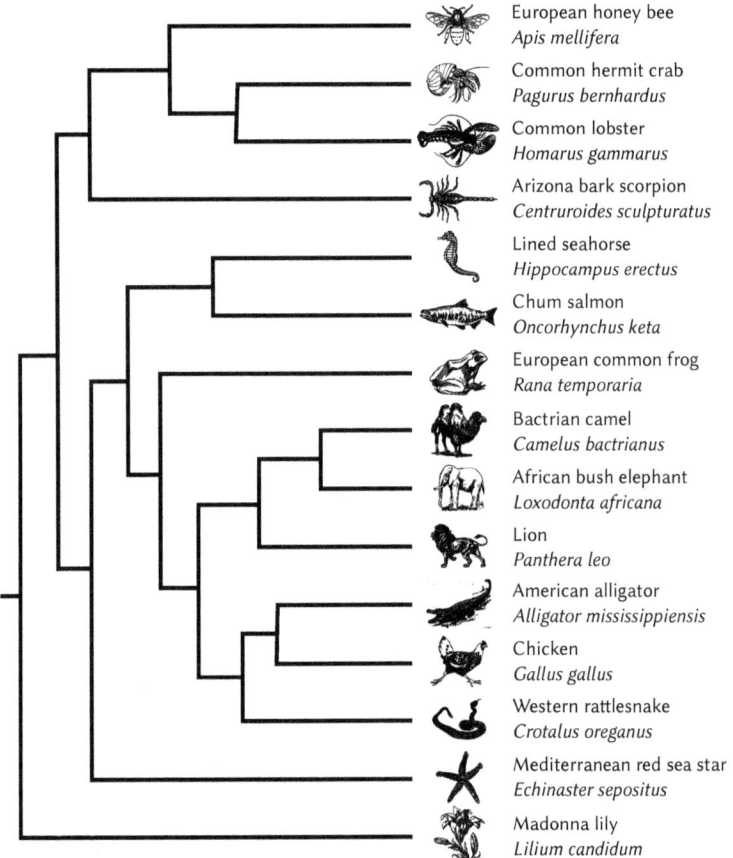

European honey bee
Apis mellifera

Common hermit crab
Pagurus bernhardus

Common lobster
Homarus gammarus

Arizona bark scorpion
Centruroides sculpturatus

Lined seahorse
Hippocampus erectus

Chum salmon
Oncorhynchus keta

European common frog
Rana temporaria

Bactrian camel
Camelus bactrianus

African bush elephant
Loxodonta africana

Lion
Panthera leo

American alligator
Alligator mississippiensis

Chicken
Gallus gallus

Western rattlesnake
Crotalus oreganus

Mediterranean red sea star
Echinaster sepositus

Madonna lily
Lilium candidum

Figure 7.1: A cladogram, showing relationships between some species. A clade consists in all species that fall below (in tree order, i.e. on the right in this picture, since the root is on the left) a branch point. For example, there is a clade that contains camels and elephants; it is a sub-clade of the clade that contains camels, elephants and lions. The smallest clade that contains both alligators and rattlesnakes also contains chickens.

For example, consider Ernie, a 33-year old silverback gorilla who lives in the Bronx zoo[10]. Ernie is the alpha male in his group, with exclusive access to the females, and thus developed the characteristic silvery mane. Then consider the former US president Bill Clinton, who also has very fine silvery hair these days. You may want to work out the smallest clade that contains both Ernie and Bill; from extensive genetic analysis, that clade would contain all humans and all gorillas, and also all chimpanzees and bonobos, because the last common ancestor of gorillas and humans also spawned chimpanzees and bonobos. However, orangutans are not from that clade. The last common ancestor of humans and orangutans is further up that tree. In that sense, gorillas like Ernie are closer cousins to humans like Bill than orangutans are; but chimpanzees are closer still. From Ernie's point of view, though, Bill and a chimpanzee are both at the same distance: they are two descendants of a cousin branch in the big family tree.

Using clades for nomenclature has interesting consequences. One of them is that there are groups which no longer exist, such as "fish" and "reptile". Indeed, reptiles would classically include all these cold rampant beasts with scales. But in cladistics, if we make a clade of reptiles, then we must include the common ancestor to all reptiles, and all its descendants. Putting crocodiles in the "reptile" category along with lizards would require going up to the common ancestor for both, and its descendants will then include all archosaurs; the *Archosauria* group is defined through some specific characteristics of the skeleton, in particular in the skull and the femur. With these morphologic criteria, archosaurs include crocodiles, but also

---

[10]In 2016. Ernie was born in a zoo in Oklahoma in 1983.

all dinosaurs, all pterosaurs[11], and, crucially, all birds. But not lizards. Therefore, under the important assumption that the shared archosaurian characteristics in the skeleton denote a common family history, if lizards and crocodiles are in the "reptile" clade, then so are birds. And that's just when trying to group lizards and crocodiles; if we want to be more inclusive and put everything terrestrial with scales in the "reptile" clade, then mammals enter the mix. So, unless we want Ernie and Bill to be reptiles, we need to abandon the idea that "reptile" has a place in phylogenetic nomenclature. On the other hand, archosaurs are a clade, dinosaurs a sub-clade of archosaurs, and birds a sub-clade of dinosaurs. We can thus rejoice in the knowledge that whenever we eat eggs, we really feast on the younglings of dinosaurs, which is a very classy way to have breakfast.

Building cladograms requires working out who is a closer cousin to whom. This can be extremely difficult and hazardous to do, especially when trying to do it with extinct species that we only know through a few fossils. A particularly vexing phenomenon is the one called *convergent evolution*: since natural selection favours traits that promote survival, it may happen that the same environment may select a similar trait in otherwise unrelated groups. A striking example is flight; in the animal kingdom, groups of animals have developed powered flight, by flapping wings, at least four times: insects, pterosaurs, birds and bats all "learned to fly" independently. Thus, presence of wings is not a proper criterion for building cladograms. The difficulty is in identifying traits that really come from genetic inheritance.

---

[11]Pterosaurs are the "flying reptiles" from the same era as the dinosaurs – but they are not dinosaurs themselves, only a related group.

Genetics and sequencing come to the rescue. As we saw previously, genetic mutations occur randomly, and most of them fall in "junk DNA" where they do not have any impact. Therefore, if we sequence the DNA of some life forms, and compare them, we should be able to understand their history. For example, if Ernie and Bill share a mutation that Siska, an orangutan born in Chester Zoo (United Kingdom) in 2015, does not have, then cladistic analysis will say that Ernie and Bill must have a common ancestor who had the mutation, and was not an ancestor to Siska. A single mutation will usually not be enough to come to any definite conclusion, because it may occur several times, or be reversed with another mutation, or occur in one chromosome in an individual who then transmits the other unmutated chromosome to some of its children. However, with billions of base pairs and a lot of mutations to work with, a clear situation often emerges from the statistics. As an added bonus, under the common assumption that mutation rates are somewhat constant over time, genetic cladograms not only offer a vision of the ancestry of species, but may also give us estimates as to when the branching occurred, by simply counting the differences in base pairs: more differences means older branching with linear proportionality[12].

Phylogenetic nomenclature makes trees. A tree sprouts branches, but there is no merging. However, it is well-known that distant cousins can marry and have a common descent; this kind of event breaks the tree-building activity. Cladograms don't tolerate reunited branches. Thus, phylogenetic nomenclature cannot operate on single individuals, but only on groups, and it

---

[12] This assumption of constant mutation rate is called a "molecular clock" that holds only as long as cell division rate is constant, which works reasonably well for mammals, but not as well for deeper time and more varied animal types. For very large cladograms, some calibration with the fossil record is needed, but *in fine* yields results compatible with known geological events.

shows group ancestry. A group is there the smallest aggregate of individuals such that cladistics still work; roughly speaking, these are the ESU, the "evolutionary significant units". ESU differ from species separated by the "reproduction barriers" in that the species are about whether individuals *could* breed together, whereas ESU are an *a posteriori* concept that tells whether they did breed together. ESU better capture evolution history than a rigid definition of what is a species.

Taxonomists are not ready to completely embrace the fluid nature of cladograms. Linnæus gave us a nice tree structure with defined ranks, so there has been some considerable effort to improve Linnæus classification with additional ranks; for instance, there is now a new rank called "tribe" between family and genus. Minerals are definitively excluded from the tree of Life. Life is separated into three domains: *Bacteria* (unicellular organisms without a defined nucleus), *Archaea* (other unicellular organisms without a defined nucleus, but a chemistry quite different from *Bacteria*), and *Eukaryota* (whose cells have a nucleus and defined chromosomes). The *Eukaryota* domain splits into kingdoms, currently at least four, for animals, plants, fungi[13], and "the rest". In cladistics there cannot be non-positive categories like "the rest", so this fourth kingdom (often called "protists") only begs for a better split. Under kingdoms, one finds phyla, a new intermediate rank. No less than 35 phyla are currently defined for animals, one of them being *Chordata* for all animals that have at some point in their development a notochord, which is like an ancestor for a spine. Vertebrates are a subphylum of *Chordata*. Vertebrates themselves split into classes; traditionally, classes would have included the problematic ones (fishes and reptiles);

---

[13]An insight from genetics: mushrooms are neither plants nor animals, but "something else".

but with phylogenetics, other classes have been defined, such as *Sauropsida* that contains both traditional reptiles and birds, but excludes some early prehistoric animals (to avoid dragging mammals in) and with ambivalent feelings toward turtles, whose exact emplacement in the tree is still a bit unclear.

Notwithstanding the resistance of Linnæan taxonomy, enhanced with extra ranks, we could say that, at this point of our story, the Epimethean archetypes are at their all-time low. Not only are they assumed to evolve, but they even lost their semantic preeminence: they are now reduced to ESU, and individuals are no longer defined as being members of a species; now, ESU are defined as groups of individuals that happen to match the evolutionary record. But a mythological titan is like a door-to-door salesman: kick him out of the door, he'll come back through the window. Just like the idea of archetypes was about to disappear for good, it was spectacularly revived.

॰ 8 ॰

In the Kitchen, with the Candlestick

Linnæus was a firm believer in a concept that he called the "economy of nature", by which he meant that the Creator had ordained all plants and animals so that everything is taken care of as in a properly running household. Throughout his works and discourses, he used several striking examples such as the swift disposal of dead animals: if an animal dies, other animals will soon swarm around it and eat it clean. As Linnæus says, if it did not work so, then we would soon be buried under corpses. He saw in nature countless interactions that, to him, expressed a well-balanced harmony:

> [...] the closer we get to know the creatures around us, the clearer is the understanding we obtain of the chain of nature, and its harmony and system, according to which all things appear to have been created.

Linnæus did notice that Mankind could *try* to alter this balance, and may have some significant success in doing so. He takes as example Frederick II "the Great", King of Prussia. Frederick II reigned from 1740 to 1786 and was renowned as an "enlightened absolutist", by which it must be understood that he paid attention to what the philosophers of his time were saying (the German version of 18th century "enlightenment" is called the *Aufklärung*); Frederick understood the importance of industry and science, and did not thought of the people as a mere fixture of the land. At the same time, Frederick was intimately convinced that he was to be obeyed in everything, and was not about to jump from interest in science to unbridled democracy. He mostly wanted to get things done, and done thoroughly. He had his quirks; he was very fond of cherries, and noticed that sparrows eat cherries. He thus decided that sparrows should not be allowed to compete with the appetite of the King of Prussia, and therefore declared that sparrows had to go. Being himself, he organised a total war on sparrows, offering a bounty of six *pfennigs* for each dead sparrow, the collection of which being managed by his efficient administration.

Frederick's endeavour turned out to be disastrous, because the sparrows, besides cherries, also ate a lot of caterpillars. The removal of most sparrows implied an abundance of caterpillars, who devastated the cherry trees. The anti-sparrow policy had to be cancelled. But even though the sparrow hunt was generalised throughout Prussia, sparrows did not disappear altogether. This is the point that Linnæus wants to make: species have been created by God himself, who is perfect and omniscient and thus knows what He does; therefore, mere human agency is not sufficient to permanently break things. In any case, without human interference, the idea that any species could go extinct by itself was unthinkable and heretic; this would be saying that God had made a mistake.

Linnæus's position is a bit weird because he had, almost under his nose, an example of *bona fide* extinction. Cows (*Bos taurus*) are a domesticated species derived from a wild species known as the aurochs (*Bos primigenius*). The aurochs looked like a black, muscular cow with longer legs, stronger neck, and massive horns. While still relatively common in Europe during Roman times, the clearing of woodland during the Middle Ages severely reduced its range and population, and by the 13th century aurochs existed only as a few groups in the forests of Poland, that they shared with the żubr (the European bison, *Bison bonasus*, that still exists today). Then something marvellous happened: the Polish kings became aware of the rarity and danger of extinction of the aurochs. As was customary in most of Medieval Europe, big game hunting was a privilege reserved to nobility; but only the King could kill an aurochs or a żubr. The King could of course, on a per case basis, grant the hunting right to anybody he wished, but in the course of several centuries, they never did, not even for their own family members. They themselves hunted the huge animals only very sparingly. This did not prevent the continuous reduction in population size. By the early 16th century, aurochs were to be found only in the forested areas near the village of Jaktorów. In the second half of the 16th century, nearby villagers were transformed into gamekeepers and were exempted of taxes, provided that they kept the aurochs safe; an inspection mandated by the King had concluded that the aurochs were suffering from competition by domesticated cattle and horses, who fed on the same grass. It was thereafter forbidden to bring any domesticated animal to the aurochs feeding places. The Polish monarchs were obviously taking the conservation of the last aurochs very seriously.

Population still declined. Another royal audit in 1602 found that there were only three males and one female left; many beasts had apparently died from an unnamed illness that was transmitted from domesticated cows. In 1619 or 1620, the last male died; its antlers were collected and set in metal rings, then sent to King Zygmunt III Waza. On the rings, the following text was engraved:

> *Horn of the last aurochs of Sochaczewski primeval forest, sent by the woiwod of the Rawski province, Stanislaw Radziejowski, the starosty of Sochaczewo, in the year 1620.*

The last female finally succumbed in 1627. Thus ended the aurochs.

The story of the aurochs shows that the concept of extinction was already well known by the time of Linnæus, thus making his reluctance to admit it kind of weird. In the specific case of the aurochs, Linnæus simply declared that domestic cattle and aurochs were only subspecies, thus there was no actual extinction of a whole species.

Linnæus's views were not out of the ordinary in the 18th century. If species were fixed, it made sense that they would also not disappear, because otherwise we would ultimately run out of species. As Linnæus's comments on the King of Prussia show, extinction was supposed not to be permitted to happen by God, but even if it really happened, then it would necessarily be through human intervention, and was bound to have negative consequences. This specific idea survived. As we saw, evolution and its genetic mechanisms forced scientists to abandon essences, fixity, and even the notion of "species". But extinction was, at the same time, awarded a strongly negative

Figure 8.1: An aurochs, with its characteristic athletic build, dark coat and curvated horns. This 19th century painting by Charles Hamilton Smith is a copy of a 16th century original, which has been lost since.

moral value: extinction is murder. Through that concept, archetypes strive: species no longer exist as templates for the living, but they reappear as "that which can be killed". To see how that concept resonates through our collective minds, we must review what is meant by extinction, and what we know about that phenomenon.

There are four main types of extinction. The first one is called pseudoextinction, or phyletic extinction. It is the simple consequence of asserting that species evolve. To take Darwin's example, consider a population of finches that is impacted by a change of environment: a volcanic eruption alters the shape of an island, its soil and exposure to winds and rain, implying a change in the available plant food. The finches live and die, and the species drifts because finches with beaks more adapted to the new food situation are more likely to have children than other finches. Thus a new species is born. The moment you say that the finches are no longer the previous species, then the old species has gone pseudoextinct (unless there were other populations of that species on other islands, of course). This is administrative renaming. This has also been called *anagenesis*. The various species that replace each other through anagenesis are sometimes said to be a single "chronospecies" that morphs over time.

The second type of extinction is when a group dies off without any descent. For example, the aforementioned volcanic eruption so altered the environment that the remaining finches don't find enough to feed, regardless of their beak shape. Or, maybe, the finches may survive on their own, but another species of bird is also present, that has a more efficient beak shape to start with, so the finches are at a competitive disadvantage. The speciation mechanism described by Darwin shows how a new species emerges out of an existing population through competition: individuals that are more apt at having offspring, have more offspring. That kind of competition can also happen between two already existing, different species that must share the same limited resource, and the consequence can be the complete disappearance of one of the groups. That's extinction.

A third type of extinction is *mass extinction*. I call it a distinct type of extinction because I am most interested in how such phenomena are perceived by human minds, and right now, mass extinction is in its own distinct category. A mass extinction is a lot of extinctions (not pseudoextinctions) happening at roughly the same time; it is a concept from palaeontology, that must be retraced within that context. The most famous mass extinction is the one that happened at the end of the Cretaceous, with all dinosaurs (except birds) vanishing in the event.

The fourth type of extinction is extinction through human action. There again, from a purely biological point of view, whether the root cause for extinction is a volcano or a farmer who fells trees, has no real significance. But in the latter case, a human is involved, so there can be a moral value attached to the action; such an extinction can be thought of as a matter of responsibility and guilt. A few hundreds of such extinctions have been reported, among them some notorious cases such as the dodo bird of Mauritius, the passenger pigeon, or the thylacine (the Tasmanian wolf). A now highly fashionable idea is that we are in the middle of another mass extinction, called the Sixth Extinction (or Holocene extinction), and it is Mankind's fault.

To study mass extinctions is to study geology. We must thus have a proper chronology to serve as reference. The Earth was formed about 4,580 million years ago; we obtained that figure from a variety of sources, notably some remarkably old and durable crystals of zircon, that require extremely high pressures and temperatures to be assembled, and then remain "closed" even if the surrounding rock gets melted. By analysing the exact proportions of the various radioactive elements imprisoned in the crystal, one can work out the crystal age with great precision. Analysis of meteorites and rocks brought back from the Moon by the Apollo missions also contributed to our understanding of the early stages of the Solar system. The Earth, like other telluric planets (Mercury, Venus, Mars), aggregated from the dusty planetary disc, and it took millions of years for a significantly massive planet to emerge; hence, there is no real unique "birth date" of Earth. Formation of Earth was gradual; by about 4,540 million years ago, a primordial Earth had formed. Then, according to the leading theory, it collided with another smaller primordial planet; most of the fragments ultimately fell back on the Earth, but some remained in orbit and aggregated to become the Moon. Needless to say, the early Earth was a rather fiery place; all these aggregation and collisions had converted gravitational energy to heat, so the Earth was initially very hot and molten.

Geologists have divided the time since Earth formation into four *eons*. The one ranging from Earth formation to (conventionally) 4,000 million years ago is called the *Hadean*[1]. It seems that in the middle of the Hadean, the conditions on Earth became somewhat compatible with life as we know it, though not with *human* life (surface temperature down to 70°C with liquid water, but no oxygen in the atmosphere). We do not know whether life appeared at that time, because we have no fossil trace whatsoever from these very early times. In any case, around 4,100 million years ago began the Late Heavy Bombardment, during which a huge number of asteroids collided with Earth and the other telluric planets. This must not be thought of as a continuous rain, because it spanned over 300 million years, which is a very long time; but it is estimated that about 40 very large impacts have occurred during the LHB. A "very large impact" here means a resulting crater which is so large that it is a called a basin, with a diameter of 1,000 kilometres or more, i.e. as large as a country like France. Even if such impacts happened only once every 8 million years, each of them would have turned the whole Earth surface into a satisfactory replica of Hell. By comparison, the Chicxulub crater, whom we will talk about for the K-Pg event, has a diameter of 180 km. Therefore, even if life appeared during the Hadean, it was likely destroyed, repeatedly, during the LHB.

The second eon is the *Archean*, from 4,000 to 2,500 million years ago. Life was definitely present on Earth since the Archean,

---

[1]Named after Hades, the Greek god of the underworld.

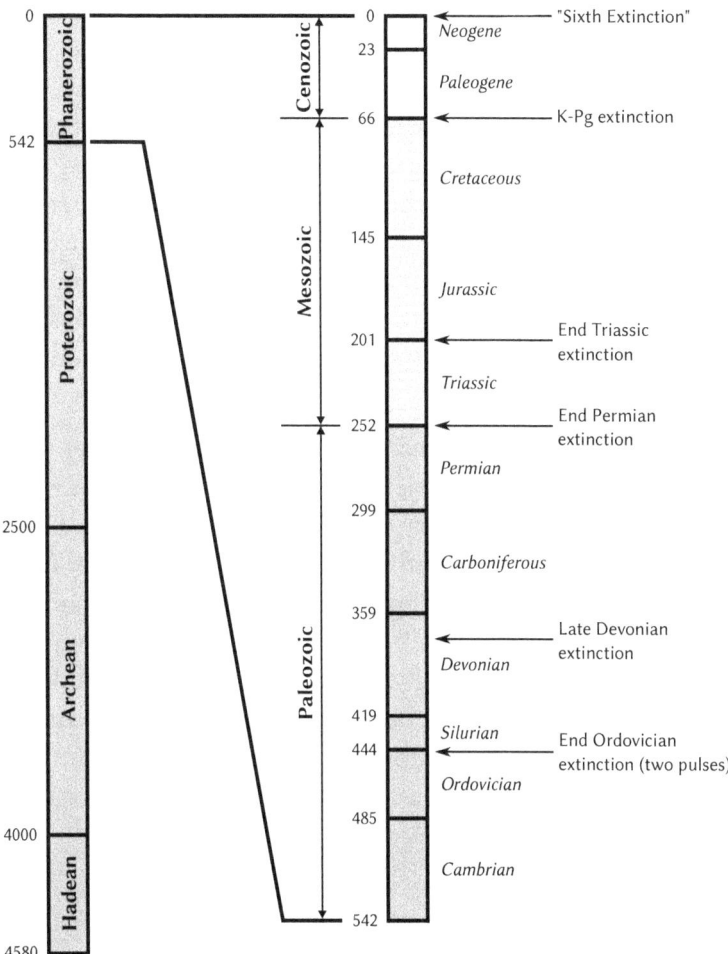

Figure 8.2: Geologic time scale, with known "big extinctions". All numbers are in millions of years. The big dinosaurs diversify and prosper throughout the Mesozoic, and vanish in the K-Pg extinction event. The first apes appear near the end of the Paleogene. The complete human history, from the first villages and beginnings of agriculture to the current day, fits in less than the last 1/1000th of the Neogene.

although the exact date is disputed. The earliest undisputed fossils date from about 2,900 million years ago. Some stromatolites were found with an age of up to 3,800 million years, thus having been formed just after the end of the LHB; stromatolites are normally the result of the trapping of sediment grains by colonies of bacteria at the surface of hot shallow waters, so these old stromatolites are often interpreted as an indirect proof of the existence of life since the early stages of the Archean. But it is possible that the old stromatolites were misinterpreted, or maybe another stromatolite formation mechanism without life is possible.

The next eon is the *Proterozoic*; the name means "earlier life". From a biologic point of view, the big event of that eon is the apparition of significant oxygen levels in the atmosphere. Organisms capable of photosynthesis, i.e. extraction of energy from sunlight, had been producing oxygen since the late Archean, but most of it chemically reacted with iron and sulfur on the surface. When all the iron was finally rusted and all the sulfur oxidized, the extra oxygen began to build up in the atmosphere, and to dissolve in significant quantities in sea water. This, in turn, allowed the evolution of organisms that breathe, converting back that oxygen into usable energy to move and reproduce.

Eons are subdivided into eras, and eras into periods. The last era of the Proterozoic is the unimaginatively named Neoproterozoic; the last two periods of the Neoproterozoic are the Cryogenian and the Ediacaran. The Cryogenian (720 to 635 million years ago) was called after its very cool climate, including two major glaciation events that may have been so severe that at some point the Earth was entirely covered in ice, sea surface waters freezing from poles to equator[2]. In the next

---

[2] This is the "Snowball Earth" theory and, of course, it is disputed, sometimes acrimoniously.

period, the Ediacaran (635 to 542 million years ago), the first complex life forms appeared; we call them "complex" because, like humans and camels, and unlike bacteria colonies, they consist in aggregations of many specialised cells, and exhibit macroscopic features. More than a hundred of distinct genera have been described, for fossils ranging from 600 million years ago to the end of the Ediacaran.

The Ediacaran life forms did not have any solid shell, so their preservation as fossils is a very rare event. This is why the first Ediacaran fossils were not found until 1868 (in Ediacara Hills, in Australia, hence the period name). Before that date, it was considered that complex life forms appeared at the start of the fourth and last eon, the *Phanerozoic*; in fact, it works in the other direction: the name "Phanerozoic" means "visible life" so it starts when fossils first appear in the geologic strata. When the Ediacaran fossils were found, the terminology was already established, so the name stuck, even though it was no longer justified.

The Phanerozoic is the current eon. Its eras are the Paleozoic (542 to 252.17 million years ago), Mesozoic (252.17 to 66 million years ago) and Cenozoic (66 million to now). To give a simple reference point: the Mesozoic is when there were big dinosaurs. The Mesozoic splits into three periods: Trias, Jurassic and Cretaceous. It may be useful to put things into perspective: complex life appeared about 600 million years ago, so the previous 2,300 to 3,200 million years were for non-complex life (bacteria and their ilk). The whole period during which interesting things happen for biology spans only the last 13% of the Earth history.

Palaeontologists have worked hard, during the last two centuries, to unearth many fossils from all periods of the Phanerozoic. Most of them are small marine invertebrates with shells. This is an important point to remember: fossilization is an extremely rare event. Most of the time, when an animal dies, its corpse is entirely eaten by other animals. Therefore, small animals, that are way more numerous than large animals, stand a better chance to be conserved. Also, fossilization normally requires water, so it happens primarily in marine environments. Finally, soft body parts are very rarely conserved, so shells and bones are the primary scientific crop of palaeontology. Most geological periods of the Phanerozoic have been identified by looking at the fossils embedded in their rocks; in particular, ammonites were very common during the Mesozoic, and their species are easily discernable through the highly varied patterns of internal shell separations in chambers.

Species identification in a fossil is a difficult art. It is very hard to know whether two roughly similar specimens represent two different species, or two individuals from the same species; we cannot apply the biological criterion or extract DNA. We can just infer things from comparative anatomy, assuming that variance within a single species more or less matches what we observe in currently existing species. As we discussed, the whole notion of species is very slippery. Thus, when discussing palaeontology, we reason mostly about genera, not species. The genus is the basic unit of what can be discerned through palaeontology.

Many genera have been identified and traced throughout the 600 million years of complex life. The fossil record is, by nature, very incomplete, but proper application of statistics can help estimate what we miss. A constant feature of fossils is that they show genera that are no longer around, so extinction is really the common lot. It is estimated than an average species lasts for 1 to 10 million years, before going extinct one way or another. Occasionally, newspapers will talk about a newly rediscovered "living fossil"; one well-known case is the coelacanth. In 1938, Marjorie Courtenay-Latimer, then curator of the East London

Museum (South Africa), was collecting specimens for her museum and had expressed her interest in unusual fish to the local fishermen; one of them, Captain Hendrik Goosen, called her to observe a large fish that turned out to be a member of an order that was believed to have gone extinct since the Cretaceous. The beast is anatomically similar to fossils that are 360 million years old. However, don't go thinking that it is the same species that survived throughout the Mesozoic and Cenozoic; the modern coelacanth is a modern species that probably appeared recently. DNA analyses show the same kind of mutations than in every other vertebrate. The really interesting part is that the fossil record clearly shows a 66 million years gap, from late Cretaceous to 1938, with no coelacanth fossil whatsoever, but the presence of the living animal forces us to think that there must have been coelacanths during all that time. This highlights the fragmentary nature of the fossil record.

Nevertheless, with thousands of genera observed from hundreds of thousands of fossils, it is possible to make statistics and try to discern patterns. During the 150 years following the death of Cuvier, in 1832, geologists and palaeontologists have been quasi unanimously uniformitarianist. For Darwin, speciation was a gradual phenomenon, so whenever two animals were considered, one being the distant ancestor of the other, all intermediate forms had to have occurred, and if we did not find them in the fossil record, then this was because that record was incomplete. However, after 150 years of scraping rocks for old bones, intermediate forms were still lacking in many cases. More disturbingly, it appeared that genus extinction was not uniformly distributed. If we plot the average rate of genus extinction over the last 600 million years, we will see big variations. There is a base level of extinction in a slightly downward sloping curve, and, occasionally, peaks that denote periods where more extinctions occur. The "Big Five" extinctions are five periods where the extinction rate is peculiarly high; they happened, respectively, at the end of the Ordovician (in two pulses, 447 and 443 million years ago), during late Devonian (375 million years ago), at the end of the Permian (252 million years ago), at the end of the Trias (201 million years ago), and at the end of the Cretaceous (66 million years ago).

It may be noticed that most big extinction events happen at the boundary between two geological periods. This is normal and expected: geological periods were identified and measured thanks to the fossils they contain, so any discontinuity in the type of fossils found in rocks was used to define boundaries between periods – and extinction events are big discontinuities.

Geology does not allow us to precisely determine the span of each event. We date fossils by accumulation of rock layers, but when it takes a hundred years to make one centimetre of rock, a simple local turbulence like the fall of a dead ten-ton dinosaur can disturb millennia worth of rocks, and it is impossible to decide when the beast died with an accuracy of less than ten thousands of years. The bottom-line is that for each extinction event, even the Big Five, we cannot know whether it happened within a week or within fifty thousands of years. Consider that the complete history of Mankind, since the building of the very first cities, fits in less than ten thousands of years. If you imagine yourself living during the extinction event, the difference in perception will be significant.

However, from a geologic point of view, these extinction events are almost instantaneous. They happen in the blink of an eye. They are definitely non uniformitarianist. They are anomalous and irksome. It was thus only natural to begin to look for a culprit, and the search concentrated on the K-Pg event (formerly known as "K-T"[3]), named after the two periods it separates (K for Cretaceous, last period of the Mesozoic, Pg for Paleogene, first period of the Cenozoic). That event was the most convenient, since it was the most recent; and, since it corresponds to the end of the big dinosaurs, it is comparatively easier to get budget for investigating the K-Pg extinction event than for any of the other four. In 1980, a team of physicists led by Luis Alvarez proposed a very catastrophist hypothesis, stating that the extinction event was caused by the impact of a huge asteroid. The evidence they offered was mostly the discovery that in rocks corresponding to the K-Pg boundary, there was a small layer of a few centimetres, found worldwide, that contains an abnormally high amount of iridium. Iridium is a heavy metal that is much rarer on Earth than gold or platinum; however, it is relatively more common in asteroids and comets. Alvarez and his team (including another Alvarez, his son Walter) thus surmised that a big asteroid had struck the Earth, triggering a huge earthquake, a heat wave that burned forests all around the world, a massive tsunami, and an "impact winter" because of all the dust thrown into the high atmosphere and blocking sunlight for several years[4]. With a big enough asteroid, the disturbance in the ecosystem is bound to imply some extinctions, starting with the biggest animals, since they are at the top of the food chain and therefore most vulnerable to any alteration.

Luis Alvarez was a well-known scientist in his field, which was particle physics; he was awarded a Nobel prize in 1968. But palaeontology is another kingdom, in which Alvarez was viewed as a trespasser – and one who marks his entry by

---

[3] In older times, eras were numbered, so the Mesozoic was the "Secondary", and it was followed by the "Tertiary", hence the "T".

[4] The "impact winter" was first popularised under the name "nuclear winter" as a plausible consequence of an all-out war between USA and the Soviet Union.

contradicting 150 years of uniformitarianism, which was felt as definitely impolite. Uniformitarianism had began to be disputed in evolution, notably with the theory of "Punctuated Equilibrium" proposed by Niles Eldredge and Stephen Jay Gould in 1972: in that theory, evolution occurs by bursts scattered in a general stability; emergence of new traits is much faster when a population is reduced, because of the statistical inertia of big groups. Nevertheless, uniformitarianism was still the dominant view, and the Alvarez' hypothesis was unlikely to be well received. It was, indeed, not well received. A heated debate began; fortunately, it triggered extra research, and controversy resulted in more data, which is, in science, a good thing. Apart from a general hostility from uniformitarianists, another candidate for a catastrophist extinction trigger emerged with the Deccan Traps, a huge accumulation of basalt, a volcanic rock, currently in India. The corresponding eruptions may have lasted for a few thousands of years, spanning over the K-Pg boundary, and the gases released in the atmosphere may have had the same "winter" effect. As a bonus, volcanic ejecta often contain substantial amounts of iridium.

Meanwhile, other proofs of a major asteroid impact began to accumulate, and in 1990, the "smoking gun" was found: the 180 km crater was identified, buried under a kilometre of Cenozoic sediments. Its centre is located near the modern village of Chicxulub, on the north of the Yucatan peninsula (Mexico). All simulations show that such an impact must have had a tremendous effect on the ecosystems of the whole planet. This does not imply, though, that the impact was the only cause; the same simulations show that the Deccan Traps also implied non-negligible consequences, so it is possible that *both* geological events concurred to drive dinosaurs to extinction; it was really bad luck to have a major group of eruptions and an asteroid impact at the same time. If they had been religious, they would have thought that the gods were really angry.

The expression "smoking gun" made the news in 1990, and most concisely characterises how the K-Pg event, and other mass extinction events, are envisioned: as crimes. When we speak and write, we use words; the ideas we express may then give an extra meaning to these words, but it also goes the other way round: the choice of words we use to formulate a concept will impart nuances to that concept. Consider then this excerpt from a 2008 article that analyses consequences of the Permian-Trias extinction event (chronologically the third of the "Big Five"):

> *The end-Permian mass extinction, 251 million years (Myr) ago, was the most devastating ecological event of all time, and it was exacerbated by two earlier events at the beginning and end of the Guadalupian, 270 and 260 Myr ago. Ecosystems were destroyed worldwide, communities were restructured and organisms were left struggling to recover. Disaster taxa, such as Lystrosaurus, insinuated themselves into almost every corner of the sparsely populated landscape in the earliest Triassic, and a quick taxonomic recovery apparently occurred on a global scale.*

Now read it again, this time looking closely at the exact words. We see that the extinction is called a "devastation". Then ecosystems are "destroyed", and the organisms were "left struggling", with the ultimate goal to achieve a "recovery". We even have an opportunist, Lystrosaurus, that benefits from the power vacuum to "insinuate" itself everywhere. All these terms convey moral values, and in particular that mass extinctions are obviously a bad thing. We are meant to empathise with the poor victims in their courageous fight to rebuild their former glory. We should boo at the Lystrosaurus who shamelessly steal the vast tracts of land that were temporarily deprived of their rightful owners.

I do not want here to berate the authors of that article; I cite their work as an example because they cannot help but using the only terminology that exists to describe the extinction phenomenon. These words shape our ideas and implicitly make us think of species as individuals with a sense of purpose. For example, "struggling for life" is the general state of all wild animals, everywhere; as Linnæus put it, the state of Nature is a general war of every being against all others. Each single animal must fight for his food and hope to avoid predators. But the meaning changes a lot when talking about a species or a genus. When we write that a *species* is struggling for recovery, what we mean is that the number of species members has considerably dropped and is not growing back with all the alacrity that would be expected... by whom, exactly?

The words we use for writing about extinction instill a number of ideas in our minds, that are not scientifically substantiated. The first one is that species and genera have a goal, namely to be "successful", with success being measured by the number of subdivisions and members of the group. A genus will be thought of as a failure if it did not have a lasting legacy, or if it contains few species. We instinctively consider species as the pawns of a game played by gods, the winner being the deity who places the most pawns everywhere, preferably with big animals. This is also apparent in usual presentations of the Mesozoic era, being the time when "dinosaurs ruled the Earth". It is highly doubtful that dinosaurs had any kind of political wisdom. Saying that a group of species is "dominant" or applies some sort of ruling is a projection from very human preoccupations. We declare that the Tyrannosaur was the king of life because he was a big apex predator, able to gnaw to death all animals that it encounters. This makes us think of absolute monarchs; the very name "tyrannosaur" means "tyrant lizard".

A second idea that is implicit in our choice of words is that a moral value is attached to the "success" of a species. It is a bad thing if a group does not strive. If its population count or variety decreases, then the group is said to suffer from loss; it struggles and tries to recover. Metaphors of sickness, decay, and death all contribute to negatively qualify such occurrences. Of course, a species cannot actually suffer; it is not a living entity with feelings. But we cannot help but empathise with the group as if it was a person, like you and me, whose business is going astray.

At this point, you should recognise the pattern. A titan is lurking nearby. Despite the patient desubstantiation of archetypes and even species since the times of Darwin, when it comes to mass extinctions, anthropomorphism and essentialism are back and at the height of their power. While our conscious analysis is about the purely statistical nature of the data, the inner mind tells us that genera are persons who are born, live and die in due order; when many die at the same time, it reeks of assassination, and such a disturbance in the normal state of Nature requires finding out a single culprit, on which we will be able to put blame. We need to find the smoking gun, because we really, really want to believe in an underlying harmony, equated with an ever expanding variety of species; a reduction of such harmony can be tolerated only if it happens because of a rare and well identified outsider whose evil deeds temporarily disrupt the steady walk of Life toward a morally superior state.

This moral undertone is what Darwin was criticising in Buffon's works. It leads us, implicitly, to lament extinction just like we would lament murder. And this in turn leads us to frame the question in terms of immediate causes and guilt: who did it? Who killed the dinosaurs? With what weapon?

And yet, for all Big Five extinctions, there was no human around the place. The last one happened 66 million years ago, and the first apes that were sufficiently upright in their walking to be called "human" appeared less than 3 million years ago. The ongoing "Sixth Extinction" is another matter.

✄ 9 ✄

The Titans Reborn

According to the *Grænlendinga saga*, in 986, a Norse merchant called Bjarni Herjólfsson was blown off his course by a storm while trying to reach Greenland from Iceland. He got sight of an unknown coast that consisted, from what he could see from his boat, of hilly woodland. Bjarni did not stop to investigate, because he was eager to reach Greenland, where his father lived. Bjarni did not arrange for further exploration; but other Greenlanders found the story interesting, because Greenland was devoid of trees. A few years later (around 1002), Leifr Eiríksson, son of Eiríkr Þorvaldsson (better known as "Erik the Red", and founder of the Greenland Norse settlement), set out to explore the new lands. He encountered a rocky and desolate land that he called *Helluland*, then a land covered with forests that he named *Markland*, and finally a coast on which he made landfall and established a small settlement. That third land was *Vinland*. The settlement lasted for a few years and was called Leifsbúðir. Later archaeological research tentatively identified Helluland, Markland and Vinland with Baffin Island, Labrador and Newfoundland, respectively; Leifsbúðir may have been the site discovered in 1960 near the small village of L'Anse aux Meadows.

In both Greenland and Newfoundland, the Norse entered in contact, then conflictual relations, with other people that they called "Skrælings", a generic and somewhat pejorative term that can be translated as "barbarians". In Newfoundland, these foreigners were the Beothuk, a tribe of people later called "Indians", "Native Americans" or "First Nations". The Beothuk, as a culture, became extinct in the early 19th century. However, DNA analysis of modern Icelanders in 2010 shows that at least one Beothuk woman had descendants that went on to live in Iceland; it is surmised that the woman was brought back to Iceland as a slave, a practice which was still ongoing in the early 11th century.

Meeting foreigners in foreign land was not a surprising concept for Norse merchants. But to 20th century palaeontologists, early presence of humans in the Americas became a puzzling problem. The Norse reached Vinland thanks to their remarkably stable boats, of a technical quality that was unmatched at that time. Even so, they did not pursue colonisation efforts for long; they abandoned Leifsbúðir after a decade; they kept making trips to Markland for two or three centuries, to gather wood that was sorely needed in Greenland. At the end of the 14th century, the Norse settlements in Greenland were themselves abandoned, and trips to Markland ceased altogether. Europeans would come back to the Americas only in 1492, by a route located much more south, and that was possible only thanks to major advances in navigation and boat construction. Long distance sea travel thus requires technology that was not available to prehistoric people.

After much research and debate, it emerged that about 13 thousand years ago, the climate was becoming warmer. From the start of the Pleistocene, 2.6 million years ago, Earth has been experiencing an "ice age", a globally cool climate characterised by an alternation of really cold and less cold periods, called glacial periods and interglacials, respectively. For the most part, the cycle length was about 41,000 years, then, in the last 800,000 years, it switched to a 100,000 year length. The leading theory for the cyclic nature is that it is related to variations in the Earth orbit and inclination of its axis; the cycles were worked out in the 1920s by Milutin Milanković. On a much larger scale, the Earth average surface temperature has been slowly but steadily decreasing over the last 40 million years, for reasons that are still explored.

One interesting theory is that it comes from plate tectonics: India was a big island that was travelling northwards, when it collided with Asia, thereby raising the Himalaya, now the highest mountain range on Earth. This exposed layers of calcareous rocks to rain. Rain is water; it may dissolve atmospheric carbon dioxide, which makes the water slightly acidic. The carbon dioxide is released when the droplets hit the ground, unless the ground consists in a type of rock that can be easily dissolved such as lime; in that case, a small part of the carbon dioxide will not be returned to the atmosphere, and will instead combine with the rock to make it soluble and ultimately end up in rivers and the sea, where small sea animals use it to build their shells. The net result is that when the Himalaya surged, more carbon dioxide was extracted from the atmosphere than previously, leading to a decreased concentration. Since carbon dioxide has a non-negligible greenhouse effect, less carbon dioxide implies a lower temperature. This is right now still an unverified theory, but it fits existing data rather neatly.

Do not get confused with time scales. Right now, over the course of the last two centuries, the Earth climate is getting warmer at a very fast rate, geologically speaking. This "global warming" has been measured by meteorologists for more than a century and with great precision; we also have good knowledge of average temperatures and atmosphere composition for the last 100,000 years, thanks to gas bubbles imprisoned in the ice sheets of Greenland and Antarctica. However, at a larger scale, when we begin to count in millions of years, the climate is *on average* cooler than during most periods of the Earth geological history. For one, we still have permanent ice sheets on large tracts of land, which is not the most common situation in the current eon. We just are experiencing a very brief warmer period. The global warming is still a matter of great interest because our human societies are impacted by what will happen within the next fifty years, not by what may happen in a million years; and the human-level consequences come more from the speed of temperature rise than from the thermal change amplitude.

During the last glacial period, smaller variations occurred, so the climate oscillated between definitely cold and awfully cold, with a local minimum about 30 thousand years ago. This implied huge ice sheets, with a depth of more than one kilometre, extended over North America and Europa. With all the water trapped in the ice, the global sea levels had also dropped by about 125 metres, turning many shallow seas into land areas; in particular, the current North Sea had become a steppe roamed by mammoths, which is why fishermen still occasionally catch mammoth bones in their nets. Crucially for our story, another of these temporary land areas was between Asia and America, where now lies the Bering strait. The area, dubbed Beringia, would have been another steppe, amenable to pedestrian crossing from Asia to North America[1]. Moreover, when the tempera-

---

[1] As we saw in a previous chapter, Beringia appeared several times during the last few million years, and allowed camels to migrate to Asia from an original American population.

ture began to rise and the ice sheet that covered most of modern Canada began to retreat, it seems that an "ice-free corridor" appeared, along the Rocky mountains, on what is now Yukon and Alberta. Thus, some human groups may have migrated through this corridor, following game and a promise of a warmer climate in the south, and finally colonised all the continent down to the Tierra del Fuego at the extreme south of South America. Since the earliest human settlement known in the Tierra del Fuego dates from about 10 thousand years ago (8000 BC), the full migratory move would have taken three thousand years, a fast but still plausible rate[2].

A migration of humans from Asia to America 13 thousand years ago is well attested in the archaeological and genetic record. The particulars, notably the famous ice-free corridors, are still disputed; it is possible that migration followed a coastal Pacific route with rudimentary canoes. The "short chronology" of the settlement of Americas is the theory that postulates that this migration was also the very first time humans set foot on either continent. The competing theory is that earlier migrations had occurred, as early as 30 or 40 thousand years ago. Various ways by which humans may have reached Americas at an early date have been proposed, with more or less plausibility. In 1947, the Norwegian explorer Thor Heyerdahl assembled a crew of six adventurers to sail across the Pacific from Peru to the Tuamotu islands, an 8,000 km trip, on board a raft assembled only from materials and techniques available to Peruvians before the Spanish conquest. The raft was called the *Kon-Tiki*, and Thor Heyerdahl wanted to prove that part of Polynesia might have

---

[2]As beautifully explained in Jared Diamond's *Guns, Germs and Steel*, North-South moves are much harder for humans than East-West moves, because they imply continuous changes of climate, hence of food sources and means to acquire them.

been colonised from South America in a westward migration, instead of the usual theory of eastward move from an initial population in South China. In 2011, a similar raft, the An-Tiki, crossed the Atlantic from Canary Islands to the Caribbean. These daring and spectacular endeavours demonstrate that prehistoric people *could* have reached America from Africa; it does not prove that they did.

Another theory, called the Solutrean Hypothesis, suggests a North Atlantic crossing by following the edge of the ice sheet, using the techniques of the Inuit people, such as canoes made of seal skin. There again, there is a definite lack of proof that such a plan ever came to fruition. The hypothesis comes from perceived similarity in stone tools between Solutrean and Clovis cultures, found in Europe and America, respectively (but with a five thousand years gap between the two). Palaeontology, so far, has failed to demonstrate that Solutrean people indeed possessed Inuit-like tools and skills to survive and strive in arctic environments.

The dispute between short and long chronology is important to our story because 13 thousand years ago, a biological event happened, called the "megafauna extinction". At that time, a large number of species of huge animals went extinct; this was particularly striking in North America. Among the species that vanished from America at the time of main human migration were the giant ground sloths, the short-faced bear, the North-American camels, some big cats such as the American lion and the saber-toothed *Smilodon*, the mastodon, several species of tapirs, the giant beaver, the armoured glyptodonts, and many others. Among big terrestrial animals, only the moose and the American bison survived; dozens of bigger species disappeared in a very short time that coincides with the arrival of humans. From these facts, a theory called the *Overkill Hypothesis* was formulated, that simply states that the newcomers hunted to death all these big animals.

Figure 9.1: Megatherium skeleton, on display in the palaeontology gallery of the *Jardin des Plantes*. The megatherium (*Megatherium americanum*) was a huge sloth that spent its days on the ground; like most big animals in the Americas, it went extinct at about the same time the first humans put foot on that continent (according to the "short chronology" hypothesis).

Why American animals would have disappeared while African megafauna, such as elephants, rhinoceros and giraffes did not, is usually explained by a question of co-evolution rate. When proto-humans first began to hunt with rudimentary tools, they caught especially apathetic animals, but those who had, out of a random mutation, an innate distrust of anything that goes on two legs would have had a better chance of survival, and thus, Darwin-style, passed on that mutation to their offspring. After some millions of years, humans had become really good hunters, a fact that all animals around them were perfectly aware of. In America, though, the newcomers arrived with efficient spear-throwers and stone points, and there was not sufficient time for local species to develop the fleeing mutation: even if an individual had the right survival behaviour, it still needed to find a prospective mate in order to have offspring, which would not happen if that prospective mate had just been barbecued. Evolution can be fast in geological terms, but it still takes a few generations to get in full swing.

Big animals are especially vulnerable to overkill, because there are not many of them to begin with, and their high content in meat makes them choice targets. Big predators also depend on the continuous availability of a suitable number of preys, so they often are the first to disappear when the ecosystem undergoes a sudden change. A similar pattern of megafauna extinction shortly following arrival of modern humans has been observed in Australia, and, at a much more recent date (around 1280 AD), in New Zealand, where the local giant flightless birds called moas disappeared within less than two centuries.

The overkill hypothesis is linked with the short chronology: in the long chronology setup, humans have been around megafauna for more than twenty thousand years, without any extinction occurring, which makes the overkill argument quite weaker. However, DNA analysis in modern inhabitants of America of native descent shows that if there were humans in America before the migration 13 thousand years ago, then these were not numerous since they were, at best, genetically absorbed in the newcomers and would have been a small minority. Therefore, even in a long chronology of the settlement of the Americas, the overkill hypothesis may still hold.

That hypothesis is still characteristic in that it is simple, even simplistic; people accept it or refute it completely, but very few are ready to accept overkill as just one factor among several causes. This is the extinction effect: since extinction is viewed as an evil anomaly, a single overarching cause is sought, that can be used to apply unambiguous blame. And, just on cue, there are humans on the crime scene. Suspicion is almost automatic.

This is a very touchy subject. To understand why, let's examine a controversy that happened in the late 1990s. In 1996, the remains of an ancient human skeleton were unearthed on a bank of the Columbia River near the city of Kennewick, in the south of the state of Washington, USA. Radiocarbon tests dated it to about nine thousand years before the present date[3]. The elders of the Native American Umatilla people then claimed that the skeleton was covered by a US Federal act called the Native American Graves Protection and Repatriation Act. The NAGPRA is part of the ongoing normalisation of relations between Native Americans and the US government, and includes provisions for returning "cultural items" to the descendants of Native Americans from which they were unlawfully taken. It is an element in the collective contrition of the American society for the historical mistreatment of the people who were already there when the Europeans began colonisation.

Under the NAGPRA, the Umatilla could request the immediate return of the skeleton to their safekeeping, and no scientific study could take place without their consent. And they were unlikely to consent, because the archaeologist who discovered

---

[3]Dates "before present" count time backward, with a starting point in 1950.

the bones, James Chatters, and another archaeologist, Douglas Owsley, considered that the bone features looked unrelated to today's Native Americans (and in particular to the Umatillas), thus warranting extra analysis of the DNA to see if there was something new to learn about migratory patterns. The Umatillas were incensed at that prospect, officially because they saw such analysis at yet another attempt of White Men at dispossessing them of their past. Less officially, the Umatillas were well aware that the legal framework on which they could claim rights on vast tracts of land was their status of "primordial inhabitants". If science turned out to reveal that Native Americans had occupied lands that already had human presence, then their legal status could be disputed. Moreover, white supremacist groups jumped on the story and began to claim that the Kennewick Man was of European origin. Tenants of the Solutrean hypothesis of settlement remained alert.

The US Army Corps of Enginneers, who oversaw the particular area where the skeleton was found, decided to bring the dispute to court. In 2004, it was ruled that there was no definite proof that the Kennewick Man was culturally linked with the Umatillas, thereby denying them the use of NAGPRA. Scientists got on with DNA analysis and other tools of their trade, and, fortunately for the Umatillas, it turned out that the skeleton was indeed related to Native Americans in general, and to the Umatillas in particular. This allowed the controversy to abate.

The Kennewick Man affair illustrates how people are ready to instrumentalize scientific results in the pursuit of any agenda, and how they like to frame societal and historical questions along moral values, in particular finding who is at fault. When Europeans began to colonise North America, settlers wanted more land, and were ready to appropriate what was, at that time, "Indian territory". Indians, on a general basis, did not have a notion of land ownership in the same way as Europeans, especially in the West where they were nomads; a nomad does not own land, he uses it temporarily. On the East coast, Indians practised agriculture with various crops such as beans or squash, but the total cultivated area was small and most of the land was forested. Forests were nobody's property, except as a loose warfare-related notion of territory, which is to be understood probabilistically: a Huron penetrating Iroquois territory did not feel at all that he was doing something *forbidden*, only *dangerous* because it increased the risk of encountering an enemy warrior.

From the British (then American) legal point of view, Indian territory was land whose entry by subjects (citizens) was forbidden because such an exclusion was part of a treaty signed with the leaders of the relevant Indian nations. A settler was barred from entering such territories not by Indian tribes, but by the White Men's law. Correspondingly, there was internal political pressure to amend such laws and open the lands for colonisation, removal of hostile indigenous people being felt as the administrative responsibility of the government. A constant rhetorical theme in the 18th and 19th century was that Native Americans did not use the land "properly" and thus did not deserve any kind of ownership on it. This is the translation of many European customary laws that state that land which has gone too long without any agricultural usage is declared "abandoned" and may lawfully become the property of whoever claims it and begins to till it.

In the 20th century, the White-dominated societies in North America began to feel pangs of guilt, and the opposite idea began to take ground. The Indian who neglects the land became the Indian who lives in harmony with the land. In 1854, Chief Seattle of the Duwamish (also known as Sealth, See-ahth, or a few other transcriptions), who lived in the modern state of Washington, pronounced a speech that purportedly contained the famous quote: "The earth does not belong to us. We belong to the earth." The speech itself was translated and written down only in 1887, and though the translator, a local doctor called Henry A. Smith, was quite careful to seek and obtain approval from the tribe elders (who did not include Seattle, who died in 1866), it is quite certain that a considerable amount of embellishment was applied to the chief's original prose. The speech then knew many variants, and ended up being one of the symbols of environmentalism. The main concept at work here is that Indians used the land in a responsible way, partaking to the gifts of Nature and respectful of all living beings. This idea has become part of the arguments by which various Native American groups have obtained and try to obtain compensations for the European colonisation.

The ecologic Indian theory is somewhat marred by the overkill hypothesis: driving dozens of species to extinction does not look like the most respectful way to live in harmony with Nature. Therefore, it is no wonder that evoking that hypothesis tends to generate acrimonious opposition. It is difficult to discuss the matter in a purely scientific way, because it is intertwined with political, emotional and moral issues.

The same issues are apparent when talking about *biodiversity*. That weird term was coined in 1985, and lately it has become a rallying cry for every politician around the world. The United Nations have declared 2011-2020 to be the "United Nations Decade on Biodiversity". Recently, the French government has created a *secrétaire d'État* (that's the rank just below "minister" in the current version of France's complicated court etiquette) dedicated to biodiversity, without any clearly defined goal as to what to do about it. But what is biodiversity?

Ontario, one of the provinces of Canada, has a "Minister of Natural Resources", who offers the following definition:

> *Ontario's biodiversity — our natural wealth of ecosystem, species and genetic diversity — has helped to shape our history, identity and economy. Biodiversity provides us with irreplaceable ecosystem services, including clean air and water, productive soils, food, fibre, timber and renewable energy. Ontario's people are healthier, and our quality of life better, because of our biodiversity.*

In other words, biodiversity is the current name of Demeter, goddess of agriculture and fertility. Any aggression against biodiversity shall be viewed as an attack against the gods, who may retaliate against Mankind in relatively unparsimonious ways, as Zeus did with his flood. Therefore, any infidel who dares decrease biodiversity (i.e. trigger extinction of a species or subspecies or any "Evolutionary Significant Unit"), or even commit blasphemy against biodiversity, shall be despatched most harshly.

It is a fact that, rationally speaking, and all other things being equal, conserving biodiversity seems like a good idea. The prosperity of Mankind, the material well-being of individuals, depends on the reliance and robustness of a subset of Nature. Alfred Henry Lewis wrote in 1906: "there are only nine meals between mankind and anarchy". In Western societies, we have become quite detached about food production, since we simply find plenty of it in any supermarket; but the continued supply of food is the result of a very complex industry. Food security was only recently achieved, even in modern countries. In Europe, in periods of peace, famine raged as recently as 1849 in Ireland, 1868 in Finland and northern Sweden. However, for all the might of agronomic science, there is still a lot that we do not completely understand about proper soil fertilisation and plant development. For example, the role of bees in pollination cannot be overstated; without bees, many crops would be unavailable or much more expensive, such as apples, cucumbers, mustard or coffee. Hence, the current decline of bee populations in both Europe and North America is a cause of worry.

Another good argument for conserving biodiversity whenever possible is that all interactions between species constitute together an enormously complex chemistry lab, that constantly tries new combinations. Many industries, in particular pharmacology, use biodiversity as a reservoir of new ideas. It has been estimated, for instance, that the discovery of antibiotics, and in particular penicillin, a production of a microscopic fungus, may have increased average life expectancy of humans by thirty years. Despite continuous advances in medicine, people still get sick and die, so it is important to keep searching for new treatments, and availability of a biodiverse ecosystem seems to be an important parameter for that search.

Despite being good arguments, these are not what drives people who are most involved in conservation. "Conservation" is the generic name for all activities that try to keep biodiversity high, so in practice this means preventing extinctions. There are many groups of researchers who go to great lengths to try to keep alive and breed the last members of some species that are on the brink of extinction, such as the Sumatran rhinoceros (about eighty individuals in the wild, a handful in zoos, and a captive breeding program that produced only four calves in two decades) or the Hawaiian crow (extinct in the wild, currently a bit more than a hundred in captivity, obtained from an all time low of nineteen individuals). It is fair to say that whatever role such beasts had on the ecosystem has already been lost – if removing them triggers a chain reaction that ends up in a catastrophe for human food production, then the catastrophe already happened. Similarly, one species of crow or of rhinoceros represents only negligible chemical complexity with regards to the millions of species of insects or plants. Big conservation efforts go to species with high symbolic value, not really to the small critters that would make more sense to preserve if the goal is to improve human well-being in the long term.

Even more extreme than conservation is de-extinction. That term designates attempts at recreating individuals from species that have gone extinct. Let's consider the aurochs. In the 1920s, two German brothers, Heinz and Lutz Heck, who were directors of two distinct zoos (in Munich and Berlin, respectively), began to crossbreed some cattle subspecies, trying to isolate traits that were deemed aurochs-like. This was before the discovery of DNA and its role, but the two brothers assumed that since cattle was domesticated aurochs, the cows had still in them the recipe for aurochsness. They produced a new subspecies, called "Heck cattle", that looks vaguely like an aurochs in body and horn shape, but is smaller than the original animal. In 1996, the *Arbeitgemeinschaft Biologischer Umweltschutz*, a German conservation group, did further crossbreeding, using Heck cattle and other varieties not used by the Heck brothers, and produced the "Taurus cattle", who are a bit taller, slender and athletic than Heck cattle, hence considered closer to the aurochs[4].

---

[4]It can be observed that the economic value of these breeds decreases as they become more aurochs-like. Indeed, domestic cattle was patiently selected to be bulkier and fatter, so that more meat is obtained; the de-extinctionists try to reverse these changes.

In 2007, a Dutch foundation called *Stichting Taurus*[5] launched in association with several universities the TaurOs Project, whose aim is again to obtain an aurochs through selective breeding, this time guided by DNA analysis: there are exploitable aurochs samples in museums, so the claim is that if one can breed a cow into having DNA that will be almost identical to that of an historical aurochs, then that cow will *be* an aurochs. From that point, it is just a small step to the idea of *modifying* the DNA directly, instead of simply looking at crossbreeding results, in order to get the expected DNA right away. This is what the *Polish Foundation for Recreating the Aurochs* (Polish acronym is "PFOT") purports to do. In simple terms, extract DNA from some preserved aurochs horn or hide, and do a straight cloning, using a denucleated cattle egg to receive the extracted DNA, and a cow to implant the embryo. In more realistic terms, start with cattle and patch the DNA piece by piece in the same way as any other GMO. It is ironic that these projects try to cancel human-induced changes in *Bos primigenius* by using the advanced genomic tools that environmentalists usually most despise and decry.

Conservation and de-extinction aim at conserving or recreating some elements of "pristine Nature". This makes sense only if one considers that human action is, by definition, unnatural. In the palaeontological record, it is clear that many species have had a big impact on their environment, starting with the first organisms that could do photosynthesis and filled the atmosphere with that well-known pollutant, oxygen. If we follow Darwin, then humans are just a kind of ape, and thus are part of the natural world and would be perfectly entitled to alter the environment. But conservation, at its core, is about denying Mankind

---

[5]"Stichting" means "foundation" in Dutch, and "Taurus" apparently stands for "Taurus", the latin name of cattle.

their membership to Nature. Humans are a foreign body whose action is, by definition, artificial, and therefore should be un-done whenever possible.

The philosophical basis for such an assertion is responsibility. Humans, alone in all biological entities on Earth, have moral notions; they know right and wrong. Because of that, they can be held accountable for their actions, and cannot just go on ex-tinguishing other species. Adam and Eve ate from the fruit that gave them the knowledge of good and evil; Prometheus stole the intelligence from the gods and gave it to Men. This concept ex-plains conservation and de-extinction as a way to repay a moral debt, to atone for the murder of species. This is why such ef-forts concentrate on big, symbolic animals like the aurochs, not on beetles or microscopic algae. This also explains why the feral camels in Australia must die: they were domesticated, making them unnatural, and therefore ineligible for conservation.

All this reasoning makes strong use of Epimetheus's figurines: we save or recreate a species, and are content with making some individuals that are similar enough to the original beast, i.e. that express the archetype with satisfactory fidelity. Prometheus's hubris is also much active: conservationists claim to restore a natural, untouched order, but in practice they are firmly in charge. This is how the inherent contradiction of repopulating a wild species through captive breeding can be solved: while captive breeding is, by definition, very artificial, it can be deemed, if applied with sufficient control, to cancel out some previous artificialness and thus recreate wildness.

In northeastern Siberia, on the banks of the Kolyma river, in an area that was basically tundra, a nature "reserve" was created in 1988, called Pleistocene Park. It is not an actual reserve in the sense of conserving an existing environment; instead, the project aims at recreating an arctic steppe such as existed during the last glacial period. This requires a lot of grazing animals, big mammals from species that roamed that steppe 30 thousand years ago, or from close enough species if de-extinction cannot produce the required specimens[6]. To some controversy about the dangers and inappropriateness of destroying a part of the "natural tundra", the project leader, Sergey Zimov, responded:

> *Tundra — that is not an ecosystem. Such systems had not existed on the planet [before the disappearance of the megafauna], and there is nothing to cherish in the tundra. Of course, it would be silly to create a desert instead of the tundra, but if the same site would evolve into a steppe, then it certainly would improve the environment. If deer, foxes, bovines were more abundant, nature would only benefit from this. And people too. However, the danger still exists, of course, you have to be very careful. If it is a revival of the steppes, then, for example, small animals are really dangerous to release without control. As for large herbivores — no danger, as they are very easy to remove again.*

This most concisely states the whole point. That specific conservation effort is about making a more interesting, "improved" Nature, under strict human control; and it is asserted that Nature will "benefit" from it in some way, thereby assuming that

---

[6]Japanese scientists from Kyoto University plan to harvest DNA from the frozen remains of woolly mammoths, found in Siberian permafrost, and to implant it in an elephant embryo, thereby recreating a complete, live mammoth. They started in 2011 and hoped to be done within six years. Other teams are also working on the subject, so there is some competition that spurs continuous efforts.

there is a notion of right or wrong that fits Nature, tundra environment being rather down that scale.

We want to become gods. Or titans.

Conclusion

Evolutionary biology, after centuries of development, has established a picture of Life as a dynamic, ever-changing structure where all individuals are, ultimately, different from each other. Nevertheless, we, humans, insist on thinking about them as broad categories, each being centred on an assumed template. We even use these categories as foundation for our moral values with regards to ecology, regardless of the difficulty to actually define them. As we saw, this theme of archetypes is very old, since it was already formalised as a creation mythology in ancient Greece, more than two millennia ago.

This quirk of our minds may come from the profound inability of human brains to deal with notions of averages. This was captured most strikingly in the famous quote attributed to Joseph Stalin: "one death is a tragedy, one million deaths is a statistic" (Stalin probably never said that, but let's not ruin a good quote with mere facts). Big numbers have the magical effect of destroying our instinctive perception of concepts. This can be easily witnessed in discussions regarding global warming: most people fail to grasp that the important word is "global", not "warming". Documentaries will show a polar bear stranded on a shrinking ice block, alone in the middle of a fully liquid sea; but that bear, regardless of its dire predicament, is not an exact assessment of global warming. What matters, for the rest of us, is not that there is less ice in one place or another; it is that there is less ice *everywhere* at the same time. You cannot "see" that in a video, except if you use satellite views that can encompass the whole Arctic in one glance. The viewer's brain can comprehend a statistical effect such as global warming only if it can be pushed as a whole into his field of vision.

The same mechanism prevents us from envisioning a crowd, or a species, as anything else than a single entity. We need to reduce the group to a symbolic individual, like an unlucky and overheated polar bear, that will serve as a metaphor. Otherwise, there are just statistics, i.e. columns of numbers that most people will find incomprehensible or just plain boring. As long as humans will keep on thinking like humans, and chances are that they will do so for still some centuries, Epimetheus's legacy will be honoured.

Acknowledgements

I would like to express my deep gratitude to Rory Alsop, who beta-tested the first version of this text, made very good comments, tracked the inevitable typographic errors, spelling mistakes and rampant gallicisms, and finally provided the foreword. May the gods look favourably upon him.

I am very grateful to Sébastien Desreux, whose keen eye for typography and thorough knowledge of the edition world have proved invaluable for the completion of this book in its current form.

Jordan Schroeder generously provided extensive advice on the self-publication process, and pointed me to the CreateSpace print-on-demand services, with which this book was processed.

M. R. O'Connor kindly read the text and suggested much needed improvements on the introduction; I hope I properly applied her suggestions and found a better, clearer wording.

The bust of Plato and the Ridinger etching depicting the garden of Eden are based on scans performed by the Wellcome Library and released under the CC-BY-4.0 license. I converted them to greyscale and slightly cropped them to better fit the pagination. The Wellcome Library keeps the copyright for the scanning effort, though the underlying source works have fallen to Public Domain (the bust photograph appeared in a 1912 book from Antal Hekler).

The picture of the Beast of Maastricht is derived (greyscale conversion and cropping) from a photograph © 2015 by Herman Pijpers, who shared it on Flickr under the CC-BY-2.0 license: https://www.flickr.com/photos/8259447@N06/24355965236

All other picture elements have been obtained from sources with an explicit notice of release to Public Domain. I list here the main online repositories, and photographer names (when known) for pictures which are not a straightforward reproduction of a previous two-dimensional work of art:

- Pixabay: https://pixabay.com/

    - The dromedaries have been photographed by Irina Mobinovyc.

- Openclipart: https://openclipart.org/

- Wikimedia Commons: https://commons.wikimedia.org/

    - The statues of Aristotle and of Buffon have been photographed by Marie-Lan Nguyen.
    - The megatherium skeleton has been photographed by Mariana Ruiz Villarreal.

Bibliography

For the redaction of this book, I have used many sources. Three decades ago, when you wanted to learn something, you went to the nearest library and tried to locate a couple of books that could contain the relevant information. Nowadays, you use the Internet, and scarcity of data has ceased to be an issue. In fact, this is now quite the opposite: for any single subject, the amount of available information is overwhelming. This changes the methodology of the scholar, amateur or professional; instead of working hard to create new information, he must now navigate in a jungle of facts to remove all the spurious data.

The problem is compounded by the inescapable observation that not everything on the Internet is true. For every piece of correct information, there exists another one which says the opposite. The search for data must then include many cross-verifications. A primary entry point is, of course, Wikipedia. As encyclopaedias go, Wikipedia is not bad; though redaction quality varies between articles, they are, on average, as readable and correct as any paper-based "established" encyclopaedia. But this is not sufficient. The real added value in a Wikipedia article is its set of external references, that must be followed and consulted. It is also quite useful in listing keywords for subsequent searches. There is an almost infinite wealth of data that just waits to be googled up, but the effort must still be made. Being able to read several languages (in my case, English and French) also helps: comparison between the different versions of the same Wikipedia article can reveal some variations which pinpoint the disputed elements.

Apart from Wikipedia, I will present here a few other sources, mainly books, that I read and found instructive. I give authors, titles and ISBN, not editors or years, because I know you will look them up on Amazon (or any other online vendor for books). The list below is merely a short selection, certainly not an exhaustive canon.

On the different modalities of belief, and how philosophers considered truth when applied to deities, read Paul Veyne, *Did the Greeks Believe in Their Myths?* (ISBN: 978-0226854342). From the same author, consider reading *When Our World Became Christian: 312 - 394* (ISBN: 978-0745644998), that deals with the transition to Christianism in the Roman Empire.

From Edward Luttwak, *The Grand Strategy of the Roman Empire* (ISBN: 978-0801821585) and *The Grand Strategy of the Byzantine Empire* (ISBN: 978-0674062078) present a panorama of the military difficulties of maintaining the cohesion of the Roman Empire.

A very influential essay by Jared Diamond, *Guns, Germs and Steel* (ISBN: 978-0393317558), is a must-read for whoever wants a first approach on the topic of domestication and co-evolution with human societies. Be sure to also read Charles Mann's *1491: New Revelations of the Americas Before Columbus* (ISBN: 978-1400032051), which explains at length how the American landscape, when reached by Europeans after Colombus, was in fact far from being a pristine wilderness. Also, Diamond adheres to the short chronology and the overkill hypothesis in the Americas, while Mann is more sceptical; comparing the two books is thus most enlightening (or at least most puzzling).

Still by Charles Mann, *1493: Uncovering the New World Columbus Created* (ISBN: 978-0307278241) follows the many consequences of the "great Columbian interchange", i.e. the transfer of species between the Old and the New Worlds after 1492.

On phylogenetics and cladograms, you could do worse than reading *The Tree of Life: A Phylogenetic Classification* (ISBN: 978-0674021839), written by Guillaume Lecointre and Hervé Le Guyader. This book contains thorough explanations on how phylogenetics work, and also many pages dedicated to the actual classification of many species.

A fascinating online resource on taxonomy is the *Tree of Life web project* (http://tolweb.org/) which is a browsable classification that aims at referencing all known species, in a cladistically correct tree. It is still expanding with inputs from hundreds of contributors.

The Great Extinctions: What Causes Them and How They Shape Life (ISBN: 978-1770853270), by Norman MacLeod, is a very nice treaty on the phenomenon of extinction, and in particular mass extinction. It is "advanced scientific vulgarisation": a clear narrative and many pictures and diagrams, but also solid science and technical details for the readers who want to explore a bit deeper.

In *Resurrection Science: Conservation, De-Extinction and the Precarious Future of Wild Things* (ISBN: 978-1137279293), M. R. O'Connor explores many aspects of conservation and de-extinction, in great details, covering not only the particulars of various projects, but also the hard questions such as what we are actually trying to conserve, and what sense it makes to breed individuals in captivity when we are interested in the wild species.

The text of this book was typed in the Markdown format. It was then converted using pandoc, and typeset by the author with LaTeX. The main text font is EB Garamond (at 12pt font size); headings use Linux Biolinum.

Photographs have been processed with GIMP.

Line-art figures have been drawn with Inkscape.